TOIYABE PATROL

Five U.S. Forest Service Summers East of the High Sierra in the 1960s

Les Joslin

Chief Thomas --
Thanks for endorsing this project!
Les Joslin

Wilderness Associates
Bend, Oregon

TOIYABE PATROL
Five U.S. Forest Service Summers
East of the High Sierra in the 1960s

Library of Congress Control Number: 2006901229

International Standard Book Number: 0-9647-167-5-5

Photographs by the author and his friends.

In the interest of full disclosure, the author notes that the front cover photograph is a combination of a 1962 photograph of the Bridgeport Ranger Station office building made by the author and a 1966 photograph of the author made by Mark Parratt.

Back cover 2005 photograph of the author by Pat Joslin.

Map by the author.

Printed and bound in the United States of America by
Maverick Publications, Inc.
P.O. Box 5007, Bend, Oregon 97708
for
Wilderness Associates
P.O. Box 5822, Bend, Oregon 97708
www.wildernessheritage.com

This book is again dedicated to
Pat, Amy, and Wendy,
who put up with my efforts to write then,
and especially to Pat who continues to do so
now the daughters are grown and pursuing
lives and careers of their own,
and to
H. Marion Hysell,
Forestry Technician, U.S. Forest Service, Retired,
who as Fire Control/Management Officer on the
Bridgeport Ranger District, Toiyabe National Forest,
for over a quarter of a century, made dozens of
young men like me—and, a long time after my time,
young women—into firefighters.

"Write what you know."
—Mark Twain

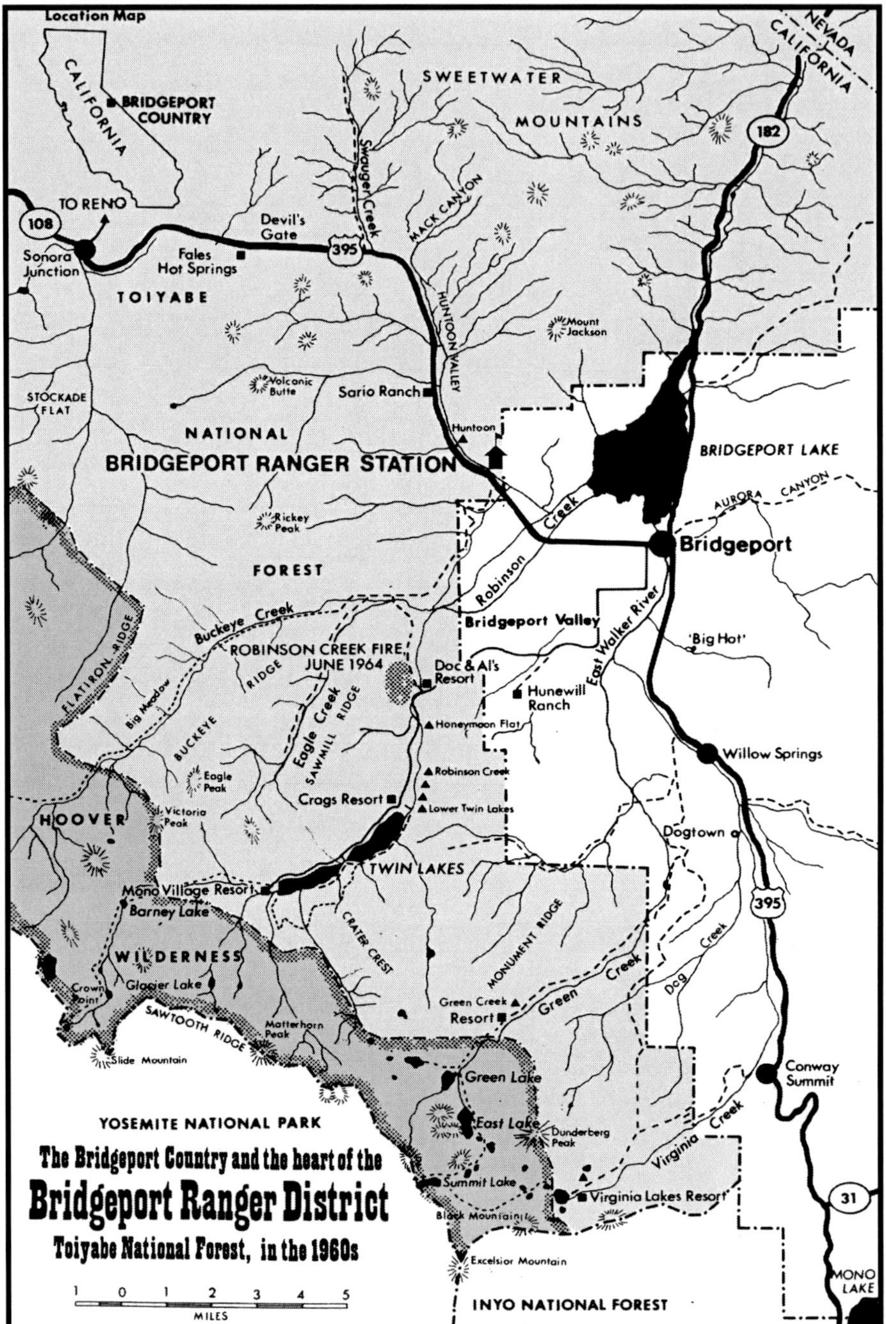

The Bridgeport Country and the heart of the
Bridgeport Ranger District
Toiyabe National Forest, in the 1960s

Chapter One: A Campfire in a Catface

I coaxed the pale green U.S. Forest Service patrol truck onto U.S. Highway 395 and drove southward. Adjusting the "squelch" knob on the radio, I pulled the mike from the dash.

"KMB-652, this is D-4-5. Radio check, over."

"D-4-5, this is 652, 10-4, 10-8," responded Pam, the district clerk at the ranger station four miles northwest of Bridgeport, California.

"Ten-four. D-4-5, 10-8." I replaced the mike and began to scan the eastern Sierra landscape. I had just left the station for another day on patrol. It was a clear, crisp mid-July morning of my fifth and final summer as a Toiyabe National Forest fire control aid. The early morning sun warmed the cab, and the air carried the exhilarating fragrance of sagebrush and pine. I smiled, rested my arm out the cab's open window, and continued my scan.

To the southwest, the granite summits of the Sierra Nevada soared to meet a brilliant blue sky. Many of those peaks, Excelsior Mountain and Dunderberg Peak rising only slightly higher than Matterhorn Peak and the jagged Sawtooth Ridge, were old friends now. Streams rising there rushed through a geologic jumble of ancient volcanic ridges and more recent glacial moraines to join near Bridgeport in the East Walker River. North of town, at the foot of the Sweetwater Mountains, the Bridgeport Reservoir briefly detained the East Walker's run to Mason Valley and the Nevada desert. Ahead, east of town, the lower Bodie Hills, wooded with gnarled and brittle pinyon and juniper, huddled beneath Potato Peak. Diamonds of dew, covering the meadows of Bridgeport Valley, sparkled in the sunlight.

Les Joslin, Bridgeport Ranger District fire prevention guard, in 1966.

At Bridgeport, the diminutive seat of Mono County, I turned toward Twin Lakes, a popular summer recreation area of campgrounds, vacation homes, and resorts nestled in a glacial canyon dominated by the steep Sawtooth Ridge escarpment. Here, where mountains and semi-desert meet, wildfire is a real threat. A total fire prevention effort included frequent patrols and intensive public contact—a big part of my job.

Following a freshman season on the fire crew, I had been promoted and appointed district fire prevention guard. To the public, I was "the ranger"—the U.S. Forest Service officer in the field. In fact, I worked under one of the district forest ranger's primary assistants, the fire control officer or FCO. In addition to all aspects of forest fire prevention and control, my job included other public relations and administrative duties. This year, my fifth on the district, I was working a full six-month appointment. I had graduated from San Jose State College in January, and was scheduled to report for active duty in the Navy the following winter. Although it has been more than a forty years since I last wore the pine tree shield for the

Toiyabe National Forest, I recall those patrols as if the last were only yesterday.

A brief trip through irrigated summer pastures brought me to rising ground and the national forest boundary. Ahead, several hundred acres of blackened brush and timber marred the eastern slope of Sawmill Ridge and reinforced my belief in my job. Two summers before, while returning from a patrol to the Virginia Lakes area, I had spotted the smoke, reported it, and raced to the scene—only to experience defeat by an inferno raging far beyond control by the district's meager assets. One day, two hundred men, and tens of thousands of dollars later, the Robinson Creek Fire of June, 1964—in reality, a relatively small project fire—was under control. But its scar would take decades to heal. If only, I mused, I could have prevented that one....

The sudden sight of white smoke snapped me back to the present. I hit the radio.

"KMB-652, D-4-5, over."

The ranger station responded.

"I have a smoke up here toward Robinson Creek Campground, but I can't see what's burning yet, over." Although the base of the thin white column was obscured from view by a low moraine, I knew it wasn't coming from the camping area.

The ranger station asked my location.

"Honeymoon Flat, just crossing the bridge, over."

"Keep us posted."

"Ten-four, D-4-5, 10-8," the ranger station signed off.

A few minutes later, I turned the patrol truck off the Twin Lakes highway and followed a little-used dirt road. It wound through blue-green sagebrush and granite boulders toward a small grove of aspen and pine. An ancient Jeffrey pine, still living though hollow from bottom to top, dominated the grove. At first, I was confused. The smoke puffing from the top of the old pine didn't make much sense. As I drew closer, the situation became clear. I decided to relieve the anxiety at the ranger station.

"What's up, D-4-5?" The FCO, attracted by my previous transmission, had come to the radio.

You wouldn't believe me if I told you. But it's no sweat."

"Try me."

"Really, it'd take too much time. I'll fill you in tonight."

"Suit yourself. Let us know if you need a hand. This is 652, 10-8."

"Thanks. I'll be off the air for a while. I'll let you know when I'm back in service. This is D-4-5, 10-7."

I halted the patrol truck behind a large, late-model station wagon bearing New York license plates. I blinked at what I saw. A short distance beyond was a half-pitched umbrella tent. Shiny new camping gear lay strewn about. Before the tent stood a woman and a boy of nine or ten, admiring a large crackling fire burning in a catface—an old open fire scar—at the base of the old pine. Rocks piled in a ring around the catface completed their "fireplace." The hollow trunk of the tree served as a chimney.

Apparently oblivious to my arrival, a stocky balding man in flowered Bermuda shorts and tennis shoes was hustling another armful of dry branches and cones toward the hungry blaze. I got out of the truck.

"Good morning, sir," I called pleasantly, smiling as I ambled toward the man.

"Huh? Oh...hi-yer," he mumbled in reply, dropping his burden. He appeared both startled and annoyed at my arrival. After gazing at me for a moment, taking in the bronze badge and green-gold shoulder insignia, he asked the question. "You a cop?"

"I'm not a policeman, sir. I'm with the Forest Service. How are you today?"

"Uh...okay, I guess...." He continued to eye me warily.

"What's going on here?" I was still smiling, but getting down to business.

"Uh...campin', yer know.... Dat against th' lawr er sumthin'?" He was at once defensive and aggressive.

"Campfires are here, sir. According to that sign you passed when you left the highway...."

"Sign?" he cut me off. "Wha' sign? I din' see no sign."

"That sign, sir." I pointed back toward the highway at a small brown signboard with a peaked roof on which a standard yellow and black Forest Service notice proclaimed NO CAMPFIRES

"I pointed...at a small brown signboard...on which a standard yellow and black Forest Service notice proclaimed NO CAMP FIRES PERMITTED HERE."

PERMITTED HERE. "Campfires are allowed only in designated campgrounds."

"Look," he flared, informing me that he was from the Borough of Queens in New York City, "I know all about fawests an' th' fawestwy! I got heah foist an' I'm campin' heah!"

"I'm sorry, sir," I countered pleasantly. "You can't camp here with that fire." He was going to be hard to win over. "That fire's dangerous. It's in violation of national forest regulations and California law." I bit my tongue to keep from remarking that I couldn't believe he'd built his fire in the old pine. "It could get away from you, spread quickly through this dry brush and timber, and do a lot of damage. A fire starting here could burn into the nearby campgrounds and endanger a lot of people." Hands on my hips, I glanced again at his Jeffrey pine fireplace and barely controlled the urge to laugh. Now, I thought, I'd seen everything. But I was young, and I hadn't. Turning back to the man, I made my best offer. "Let me help you put this fire out and find a place in the campground...."

His further protests, including comments about "bureaucwats" and the United States still being a "fwee countwy," evoked my patient restatement of the wildfire danger. The Robinson Creek burn, clearly visible less than a mile to the north, provided useful graphic support. I avoided suggesting that, if necessary, the judge almost certainly would see things my way.

"Well, awight," he finally relented. "You cops always win." I ignored the "cops" remark. "But I ain' got no wadah!" That didn't surprise me.

"I have," I said, swinging onto the rear of the patrol rig to start the pumper. For a moment, I considered handing my friend from Queens a fire tool and soliciting his assistance, but then thought better of it. A shovel or pulaski in the hands of an inexperienced person could be dangerous.

While this particular situation was unique, several seasons of patrol experience had taught me not to be surprised by what some national forest visitors may do. Many of them are, after all, in an unfamiliar environment. This time, however, I was in for a big surprise. It came as I began extinguishing the fire.

Throughout our parley, the man's wife and son had stood by as quiet observers. But suddenly, apparently not comprehending what was happening or why, the little boy bolted toward me, fists flailing as he shrilled. "Father! Father! He's ruining our beautiful campfire!" Never before had I been attacked by a child. I shot a desperate glance at the man. He, too, seemed genuinely surprised. Finally, he restrained his son and hustled him off to help pack the station wagon.

I was still mopping up when, without another word, the family roared off in a cloud of dust. Issuing the man a citation—a rarely-used last resort—would have served no useful purpose. He was out of his element, and needed education, not embitterment. I hoped my explanation, and suggestion of a camping place, had helped.

Thirty minutes later, certain the fire was out, I called the ranger station to report in and continued the patrol.

The encounter with the New York family had again set me thinking about myself and my job. After several seasons at it, I wondered, just how was I doing? It didn't seem enough to be told I was doing a good job. I had questions that wouldn't answer. Was my approach to each situation a good one? Did each face-to-face encounter involve the right mix of diplomacy, empathy, and helpfulness? Was my image one of fairness and competence? Just what kind of impression did I leave with the public? It seemed I was in a position to make good friends or bad enemies for the Forest Service on a daily basis. Sure, I was just one patrolman on one of maybe 700 ranger districts on about 150 national forests spread across thirty-eight states. But I met hundreds of people each week. Did I worry too much about such things? No, I decided, I didn't. My job was more than preventing and suppressing forest fires. It was also to win friends for and understanding of the Forest Service and its endeavors. The way I did my job helped mold public opinion of the organization I had always wanted to serve.

That's why I had to be careful. It's all too easy for someone whose job involves daily contact with a variety of publics—in my case, a wide range of national forest visitors and users—to forget these things, to become indifferent toward or even disdainful of the publics he serves. Worse yet, I knew, would be to ape the "badge-heavy" cop type or act a "big shot" around these publics. After all, these publics —visitors and users—are the citizen-owners of the National Forest System. It's "national forest land" that belongs to the people of the United States and is administered for them by the Forest Service, *not* "Forest Service land." A small point? Not at all! It's an all-important distinction that informs—or should inform—the perceptions of both the public servants and the publics they serve of their respective roles, responsibilities, and prerogatives regarding the national forests and each other. The Forest Service's 1905 *Use Book* for early forest rangers had said it best.

> The administration of forest reserves is not for the benefit of the Government, but of the people....Forest officers, therefore, are servants of the people. They must obey instructions and enforce the regulations for the protection of the reserves without fear or favor.... They must answer all inquiries... fully and cheerfully, and be...prompt and courteous.... They must make every effort to prevent the misunderstanding and violation of reserve regulations by giving information freely and fully. The object should be to prevent mistakes rather than punish them. Information should be given tactfully, by advice, and not by offensive warnings.

Gifford Pinchot, founder and first chief of the U.S. Forest Service.

"Sound doctrine," Gifford Pinchot, founder and first Chief of the Forest Service, wrote in his 1946 autobiography, "and make no mistake."

Even now, when I write about some of the situations in which I dealt with those publics, I hope not to seem insensitive to the forest visitors and users involved. Most of the people I met on patrol—and I met tens, if not hundreds, every day—were good citizens of the outdoors. They knew and understood the few rules and regulations required for their own safety and for protection of the environment. I enjoyed serving them. Some were in need of information, which I cheerfully provided. Others required some form of administrative assistance, often issuance of a campfire permit—an outstanding public contact and educational device when properly dispensed. Still others, frequently elderly retirees, just wanted to talk. That, as I saw it, was part of the job, too, and often a most interesting part. Some of those old-timers had really been around.

Although most of the people I met were middle-class California suburbanites, visitors to the Toiyabe National Forest came from all parts of the nation and the world. Contact with them, I am convinced, added immeasurably to my social education. All deserved the best and friendliest service I could provide. I learned then, and believe still, that *public service* means *just that.*

Matterhorn Peak and the Sawtooth Ridge loom over Horsetail Falls in the Hoover Wilderness, Toiyabe National Forest.

Chapter Two: The Summer of '62

When the dust cleared, I stood alone along U.S. Highway 395, squinting after the Greyhound bus rapidly disappearing to the southeast. A broad, grassy valley, rimmed by the blue-green of sagebrush-covered slopes studded with pinyon and juniper, shimmered in the brilliant sunlight of a June afternoon. A light, fresh breeze carried the scents of grass, sagebrush, and pine. An occasional unfamiliar bird call broke the silence. Across the highway, a small brown sign promised I would find the Bridgeport Ranger Station just beyond a slight rise. I shouldered my seabag and crossed the highway to follow an unpaved road. I was stiff and tired, and felt grimy.

Dad had put me on the bus in Oakland just after midnight, and for the first half of the trip I had dozed fitfully and wandered aimlessly around bus stations nasally announced as "Vallejo, Sacramento...." A spectacular Sierra sunrise at Donner Pass had thrilled me, and I was wide awake for the rest of the trip.

I had changed busses in Reno. Fearful that my seabag would stay on the bus bound for "Winnemucca, Battle Mountain, Elko, Wells, and points east," I had skipped breakfast to hang around the baggage area to make sure it got on the southbound Los Angeles bus. It did. A late morning sandwich and Coke in Carson City doubled as breakfast and lunch. As the bus pulled out of Nevada's capital, I asked the driver to let me off at the ranger station north of Bridgeport. He grinned, and assured me he would.

As the bus sped southward, I searched the unfamiliar landscape for evidence of the Toiyabe National Forest. First, just north of the Califor-

"I searched the unfamiliar landscape for evidence of the Toiyabe National Forest...a shield-shaped national forest boundary sign."

nia line and south of Topaz Lake, was a small guard station. Then, after the settlements of Coleville and Walker, a shield-shaped national forest boundary sign, and the bus was engulfed by the wild beauty of the West Walker River canyon. Another small guard station flashed by as the bus climbed toward Devil's Gate Summit that, although I didn't know it at the time, marked the boundary of the Bridgeport Ranger District. Perhaps ten minutes later, the bus slowed at the mouth of a shallow canyon, and the driver called "Still want to get out at the Bridgeport Ranger Station?" The bus stopped. Tossing my sea bag ahead of me, I stumbled off the bus.

About fifty yards along the dirt road, I topped the rise and saw it. The ranger station was a cluster of half a dozen green-roofed white frame buildings nestled in a grove of poplars, cottonwoods, and willows. Except for a green Forest Service pickup and a gray government Jeep wagon dating from the 1950s, the scene was more of the 1920s or 1930s than 1962. And, to my way of thinking, more of a ranch than a ranger station. Gravel crunched beneath my boots as I

The Bridgeport Ranger Station looked like "more of a ranch than a ranger station" to new arrival Les Joslin in 1962.

strode toward a building marked "office" and flanked by an antenna. My long-awaited Forest Service career was about to begin.

A forestry aid named Dick was manning the office. He welcomed me, told me the district ranger was in Bridgeport where his wife was about to have a baby, and pointed me toward the bunkhouse.

The place looked pretty primitive. The bunkhouse, behind the office and the ranger's residence, was three end-on-end rooms with a door at either end. The first room was furnished with four metal-sprung cots, a couple chairs, and a heater. I dumped my seabag onto what appeared to be the vacant cot and entered the small middle room. Here were a deep sink and mirror, a shower, and much to my relief because I had noticed an outhouse out back, a flush toilet. The third room, the same size as the first, was a kitchen with a big wooden table and several chairs. A black iron woodstove edged out a small gas range for antiquity. There was a sink with running water, topped by cupboards. Contrasting with the white walls, a red linoleum floor sagged toward the center. I learned later that previous residents had included Civilian Conservation Corps enrollees in the 1930s followed by World War II conscientious objectors of the Civilian Public Service Program assigned to forestry work.

I took a shower, put on appropriately faded clean Levi's and a sweatshirt, and went outside to get the lay of the land.

In addition to the office, residence, and bunkhouse, there were a twin to the bunkhouse, two large shop-garage-storage buildings, a small gas house, and a house trailer. Back up the road that led to U.S. 395 was a small weather station.

About six o'clock, two dusty Forest Service pickups rolled into the yard. A short time later, Dick and two others, who introduced themselves as Pete and John, ambled into the bunkhouse. All three were Nevadans, and full of good-natured jokes about Californians. Dick and Pete, both second-year men on the district, also regaled me with stories about the district's fire control officer, one H. Marion Hysell, who would return from a "bug job" in his native Wyoming in a couple weeks and who would be my—our—immediate boss. They invited me to ride into town with them.

On the way into Bridgeport that evening, Dick also informed me that I would start work in the morning. As a fire control aid (GS-3), I would earn $1.83 an hour.

I'd spent Christmas break of my freshman year at the University of California, Berkeley, studying for final exams, writing an Economics 1A term paper, and typing out dozens of U.S. Government employment applications—the old Standard Form 57—for summer Forest Service jobs. I passed the finals, got an "A-" on the term paper, but heard nothing from the applications until one April evening when Mr. Paul Casamajor of the School of Forestry faculty telephoned and invited me to his Mulford Hall office to talk about jobs. Some of my applications had been referred to him, and he had three "student requisition" jobs to offer me. The first was a choker-setter job with a logging outfit in Northern California. That was out! I wanted to be a ranger, not a logger! Luckily, the other two were Forest Service opportunities, one for a recreation aid on the Wasatch National Forest in Utah and one for a fire control aid on the Toiyabe National Forest.

Bridgeport District Forest Ranger Bob Hoag in Buckeye Canyon in 1962.

Trying hard not to betray my excitement, I picked the fire job. "You'll like Bridgeport," Mr. Casamajor smiled. "The last time I was there, sheep were being trailed down the main street."

There weren't any sheep in town that June 1962 evening of my first visit to Bridgeport and its general store. But there was a strange sense I'd been there before, though I knew I hadn't. Was it possible I'd seen the town in the film noir masterpiece *Out of the Past* – starring Robert Mitchum, Jane Greer, and Kirk Douglas filmed there in October 1946 – on television? That's my only explanation.

District Forest Ranger Robert F. Hoag, Jr., a World War II veteran and Syracuse University forester who had rangered a couple districts on the Ashley and Wasatch national forests before coming to Bridgeport in 1959, knew how to get summer crews—and greenhorns—lined out. Until the fire season began, I found out my first morning on the job, the three-man fire crew would maintain trails, clean campgrounds, and do project work. This would help everyone get in shape and new men like me learn the country. Once the fire

"About half the days remaining that June, the fire crew repaired switchbacks and replaced water bars on trails...and cleaned outhouses and collected garbage in the campgrounds."

danger increased, our crew would have to stay closer to the ranger station.

As a first-year man and an outlander, I had a lot to learn about the country, the organization and the forest, my job and myself. About the best thing I had going for me was my all-consuming interest. I had wanted to be a forest ranger since grade school when Smokey Bear was my hero. In junior high school, while my friends read Hardy Boys mysteries and Black Stallion adventures, I read Montgomery M. Atwater's novels about college students working summers for the Forest Service as lookouts, smokechasers, and smokejumpers. In high school, I read Gifford Pinchot's autobiography, *Breaking New Ground,* and every other book on forestry I could get from the library or the Government Printing Office. Also, I had just earned an "A" in Forestry 1 at Berkeley—the only "A" of my freshman year. I was a greenhorn, sure, but enthusiastic and informed, eager to work and ready to learn.

I worked.

About half the days remaining that June, the fire crew repaired switchbacks and replaced water bars on the Barney Lake and Horse Creek-Cattle Creek Loop trails. Most other days were dedicated to cleaning the outhouses and collecting the garbage in the lower elevation campgrounds at Twin Lakes and Green Creek already in use. The district had no garbage truck, so campground refuse was hauled in the range conservationist's horserack pickup. This variation on the "multiple use" theme required cleaning the truck before as well as after hauling garbage. Dick and I spent a day preparing a higher elevation campground at the Virginia Lakes for use—digging a cache of garbage cans out of a snowbank and distributing them, righting picnic tables overturned to protect them from the snow, and cleaning out the snowed-in outhouses. A couple days were spent maintaining fire equipment and tools. As I worked, mountain muscle replaced campus flab.

I learned.

All that field work helped me begin to learn the country. At about half a million acres sprawled from the Sierra crest to the Nevada desert, the Bridgeport Ranger District was a big piece of country—

"At about half a million acres sprawled from the Sierra crest to the Nevada desert, the Bridgeport Ranger District was a big piece of country."

big enough to have several geographies and climates. On the west, the district rises above timberline to climax in the lofty serrated summits of the Sawtooth Ridge and other peaks in the eleven- to twelve-thousand foot range that mark its border with the western slope wilderness of Yosemite National Park and the Stanislaus National Forest. East of this crest, an alpine region of lesser peaks and glacier-gouged canyons and lakes merges with complex subalpine forests dominated by lodgepole pine and firs. At lower elevations, the Jeffrey pine forests of both the eastern Sierra and the adjacent Sweetwater Mountains overlook the still lower pinyon and juniper woodland which, in turn, yields to the arid vastness of the Great Basin sagebrush scrub association. Countless small tributaries drain this high country into larger creeks—Buckeye, Robinson, Green, and Virginia, often fringed by quaking aspen—that flow into the East Walker River. This river tumbles northerly between the Sweetwaters and the Bodie Hills and around the Pine Grove Hills to its confluence with the West Walker River that, in turn, flows on to Nevada's saline Walker

Lake. I soon realized that the district's main operations—recreation, watershed, and range management—reflected this geography.

At the center of this country, in a valley by the same name, is Bridgeport, then a town of about 350 on the East Walker River that had been settled for just over a hundred years. That settlement, of course, had followed several decades of exploration and had been spurred by mining. Among the first Euro-Americans to pass through the eastern Sierra country were parties led by Jedediah Smith in 1827 and Joseph Walker in 1833. But the Bridgeport Valley itself seems to have been "discovered" by Captain John C. Fremont, U.S. Army Corps of Topographic Engineers, in January, 1844. Against orders to return east after reconnoitering the Oregon Trail, Fremont led his party—including scouts Christopher "Kit" Carson and Thomas "Broken Hand" Fitzpatrick—southward east of the Cascade and Sierra Nevada ranges. This misguided and almost fatal wandering, justified by Fremont as a search for the phantom San Buenaventura River that by then he must have known didn't exist, trapped the expedition east of the snowy Sierra. On January 25, after turning westward at Walker Lake, the desperate party passed through Bridgeport Valley and turned northward through Devil's Gate to find a way over the Sierra to Sutter's Fort and safety.

In 1857, the Gold Rush reached the Bridgeport country when miners founded the little settlement of Dogtown on Virginia Creek. But the "big strikes" were made to the south and east, and by 1860 settlers in and around Big Meadows—Bridgeport's original name—were supplying timber, beef, and whatever other agricultural products the climate would permit to the miners at Aurora and Bodie. These early lumbermen and ranchers bequeathed their names to many of the area's peaks, creeks, and meadows. Somehow, in the autumn of 1863, Big Meadows became Bridgeport. The next year, ninety-eight years before I arrived, Bridgeport became Mono County's seat when the California-Nevada border survey discovered the county seat of Aurora was really a few miles inside the new state of Nevada. The handsome white Mono County courthouse, built in 1880, remained Bridgeport's centerpiece. A ranching and timber economy,

"The handsome white Mono County courthouse, built in 1880, remained Bridgeport's centerpiece."

eventually supplemented by outdoor recreation which boomed after World War II, had supported the country since. And the lands of the Toiyabe National Forest were the basis of that economy.

"The most urgent need of the early settlers in Big Meadows was for lumber," I read in local historian Ella Cain's 1962 book, *The Story of Early Mono County*, then hot off the press. "There was abundant timber in the surrounding mountains and several small mills driven by steam were quickly built and put in operation." These mills were at Twin Lakes, on Robinson Creek, and in Buckeye and Yaney canyons—all places I would work at and know well. By 1880, these mills were "running at full blast" to help meet "the crying need for lumber, timber, and fuel for the Bodie mines" that supported that teeming camp of ten thousand souls. Many of the Bridgeport country's best stands of timber, on the Bridgeport Valley's western and northern slopes, had been cut to support the mines by the time the first Forest Service ranger arrived in 1906 and the Mono National Forest—later split between the Toiyabe and Inyo national forests—was set up in 1908 from lands previously in the Sierra and Stanislaus forest reserves.

The site for the Bridgeport Ranger Station, four miles northwest of town, was approved in 1907 while Thomas C. West was ranger. Rangers named Fulton and Tyler worked out of the station off and on for the next decade or so. In 1919, Ranger Henry W. Atcheson of the adjacent Sweetwater Ranger District took charge of the Bridgeport Ranger District and served as its ranger until 1941. Ranger Atcheson spent a dozen years improving the station's structures and grounds. He and a crew built a new warehouse in 1931. But, in 1933 and 1934, the older buildings were replaced by a new office and residence, and a second warehouse was built—all with great Depression-New Deal era funding. These structures were the heart of the ranger station at which five subsequent district rangers lived and worked before Ranger Hoag took over in 1959 and I arrived three years later.

I had known before I arrived that the Toiyabe was the largest national forest in the lower forty-eight. With over three million acres in its Sierra, Central Nevada, and Mount Charleston divisions, it was divided into eight ranger districts under a forest supervisor, Mr. Ivan Sack, in Reno. Actually, only the Central Nevada Division—with districts headquartered in Austin, Fallon, and Tonopah—had to do with the Toiyabe Range for which the forest was named. The name "Toiyabe" was Paiute for "black mountains," inspired by the black hue the range's cover of pinyon and juniper afforded at a distance. Further south, the Las Vegas Ranger District administered the forest's lands centered on Mount Charleston north of the famous gambling city. Of the four Sierra Division districts—headquartered in Carson City, Minden, and Wellington in Nevada, and at Bridgeport—the three southernmost had been part of the former Mono National Forest until transferred to the Toiyabe during a 1945 national forest reorganization.

Forest Supervisor Sack's small headquarters staff helped his district rangers run their districts. In 1962, as I recall, this staff included a deputy supervisor, William O. "Bill" Deshler, staff officers for major national forest resource uses, and administrative and technical assistants. The later legendary Archie Murchie, whose 1929 to 1965 career in the Forest Service was immortalized in Professor R.T. King's

"Forest Supervisor Sack's...staff helped his district rangers..."
District Ranger Bob Hoag and Deputy Forest Supervisor Bill Deshler work on plans at the Bridgeport Ranger Station in 1962.

1991 book *Free Life of a Ranger*, was the range staff officer. Lyle Smith, who had been Bridgeport District ranger from 1947 to 1951, was the recreation and lands staff officer. Frank Dunning was fire staff officer. Harry Tullis was forest engineer.

Of course, I had a pretty good idea what the Forest Service was about. I knew it had been established in 1905 to manage the forest reserves—soon renamed national forests—to provide for "the greatest good of the greatest number in the long run" in Gifford Pinchot's words. And I knew that the Forest Service was striving to manage these lands—some 150 national forests comprising over 180 million acres—and their primary resources—timber, water, forage, wildlife, and recreation—under a multiple-use sustained-yield concept to maximize their benefit to the nation. It didn't take me long to realize that the Bridgeport Ranger District was a big piece of country for Ranger Hoag and his small permanent staff—a range conservationist, a forestry technician who doubled as fire control officer, and a district clerk—to manage according to these guidelines.

I knew, also, that the Forest Service cooperated with the states and private forest owners and conducted forest and range research. I soon found out that, as a "Cal Co-op" firefighter whose job was funded by the California Division of Forestry (CDF) as well as the Forest Service, I was part of that cooperation-with-the-states effort. California didn't have its own fire crews in that part of the state, so Forest Service crews took action on "state fires" as well as fires on national forest land. Additionally, since the nearest Bureau of Land Management (BLM) fire crew was in Carson City, eighty-five miles north of Bridgeport, we also suppressed fires on vast public lands in our vicinity managed by that U.S. Department of the Interior agency.

That was a lot of country for Ranger Hoag's small fire organization—a fire control officer, a fire prevention guard, and a three-man fire crew backed up by a three-man trail crew and a few other district personnel—to cover. But cover it we did.

I chased my first smoke on June 23, along with the district's veteran forestry aid Maurice Crawford and the other two fire crewmen, before the FCO returned from Wyoming. About forty miles from the ranger station, it was a lightning-caused BLM fire burning in the open pinyon-juniper woodland of the Mono Basin. Not long after we arrived and started building a fireline, an air tanker from Carson City dropped a load of pink slurry on the fire and the firefighters. Since the district lacked a pumper truck, that air drop proved a great supplement to the milk cans of water we hauled in the pickup. The pink slurry also proved very difficult to wash off the pickup! A pumper crew from the Mono Lake Ranger District, Inyo National Forest, helped us mop up.

It really wasn't much of a fire. Less than an acre of pinyon and juniper, as I recall. But it was a thrilling first visit to the Mono Basin—John Muir's "country of wonderful contrasts" I would write a master's degree thesis about some twenty-two years later—that holds me in thrall to this day. I guess I'd seen it on the map, but I really wasn't ready for what I saw as the pickup topped Conway Summit on the way to that fire. The brilliant blue expanse of Mono Lake—the "solemn, silent, sailless sea" of Mark Twain's 1872 classic *Roughing It*—with its black and white islands, dominated a Mono Basin golden in the late afternoon sun. "Hot deserts bordered by snow-laden mountains, cinders and ashes scattered on glacier-polished pavements, frost and fire working together in the making of beauty," Muir had written in *My First Summer in the Sierra* of the scene I beheld that June afternoon of my first summer in the Sierra. And even his prose fell far short.

We got back to Bridgeport just in time for the two-dollar fire meal Uncle Sam bought his after-hours firefighters in those days. That night it was a chicken fried steak at the Trails Restaurant for $1.80. That left just enough change for coffee and pie.

One morning, soon after that first fire, Ranger Hoag assembled the crew in the yard to line us out for the next several days. Mrs. Hoag, he explained, had her hands full with their new baby, and could no longer be the district clerk. "Can any of you guys type?" he asked.

"I can," I responded, before realizing what I was doing.

"Good. You are now the district clerk." And that was that. In addition to my fire crew duties, I would staff the office from time to time. I would type and file, greet visitors, take and report daily fire weather readings, and maintain radio contact with crews in the field. At fifteen by twenty-four feet, it was a crowded but comfortable one-room office. A big oak desk facing the front door supported the district's single big, black Underwood manual typewriter. A smaller oak desk, a couple oak chairs, a bank of oak file cabinets, and an oak bookcase crammed with green-bound Forest Service manuals completed the furnishings. Framed district maps covered the wood-paneled walls. Windows and bare bulbs provided light.

That evening, Ranger Hoag called me into the office. "Here," he said, handing me a small object. "Wear this so people will know who you work for." It was a bronze Forest Service badge.

"I'm gonna work the holy hell outta you," Marion Hysell greeted me when I was introduced to him on his first morning back from the Wyoming bug job. His handshake was firm as he appraised his new firefighter.

"That's what I came for," I managed in return, half wondering if what I'd been doing the last couple weeks had amounted to nothing more than resting up for the FCO's return. Then the twinkle in his eyes and his good-natured grin allayed my apprehensions.

The fire control officer and his fire crew spent the last day of June and the first three days of July constructing a new fire weather station and installing instruments. In addition to my share of digging fencepost and anemometer post holes, carpentry, and painting, I got to affix the anemometer atop its pole. Daily afternoon readings—temperature, relative humidity, wind velocity, precipitation, and fuel moisture—taken at the old weather station, and soon to be taken at the new one, were plugged into the wildland fire danger rating system to help determine needed levels of fire prevention and control efforts.

There's a lot to know about fighting forest and range fires safely and effectively. The forest had no official "fire school" that year, but Marion—who insisted on being called by his first name when I addressed him as "Mr. Hysell" as I had been brought up to address my seniors—included informal training sessions in the daily routine. We learned from everything we did.

On the theoretical side, I learned that the Sierra Nevada itself helps create the Bridgeport country's wildfire danger. Moisture is squeezed from the eastbound Pacific air as it rises over the verdant western slope of the range. The dry air that flows over the steep eastern face, warming as it descends, sucks the moisture from the vegetation and creates a wildland fire hazard to be reckoned with. The risk of ignition by summer lightning storms and human carelessness makes wildfires inevitable.

On the practical side, I learned how to find and fight fires in that big country. The first thing I learned was that fire control on the Bridgeport District in 1962 was pretty primitive. The district had no lookouts and only limited access to aircraft, and most fires were detected when one of us or a member of the public sighted a smoke. The district had no pumper truck, so whatever water we put on fires

The smokechaser pack carried by Bridgeport Ranger District firefighters in the 1960s contained a pulaski, shovel, canteen, rations, headlight for hardhat, map, and other useful odds and ends.

we transported in milk cans on a pickup, if we could drive to the fire, and in five-gallon rubber water packs on our backs, if we couldn't. Since most of the district's fires were isolated ones ignited by lightning, our primary piece of firefighting gear was the smokechaser pack, carried on the back of a firefighter, that contained a pulaski, shovel, headlamp, canteen, rations, map, and other useful odds and ends. When a smoke was reported, two firefighters equipped with these packs and a two-way radio drove as close to the fire as they could, got out, and started walking. Once they found the fire and put it out, they walked out. And so it was that my early fire fighting education emphasized the use of hand tools and little or no water.

On the morning of July 4—it was a legal holiday and we were on fire standby duty—I helped Marion, Maurice, and Pete, the district fire prevention guard, build a float on the Chevy flatbed for Bridgeport's annual Fourth of July parade. Steve, a University of Utah medical student on the trail crew, donned the Smokey Bear suit sent from the regional office in Ogden and rode the float. It was a beautiful day, the parade was a great success, and the rest of my day in the ranger station office was busy with visitors but otherwise uneventful.

That first summer afforded several opportunities to put my new fire fighting skills to use. The most challenging of these was a fire high in the crags above Devil's Gate Summit reported one July evening by a passing motorist. Dick and I were dispatched to the fire in a pickup into which we'd loaded two smokechaser packs and two water packs. It didn't take us long to spot the smoke and to size up the climb ahead of us. We parked the truck on the south side of U.S. Highway 395, put on our hard hats, strapped on our smokechaser packs, and began the climb—which often involved crawling up steep sliderock slopes—toward the fire. I soon realized how well six weeks of hard work had conditioned me.

About forty minutes of climbing later we reached the fire. It wasn't much as fires go—a large Jeffrey pine snag, a few smaller pines and junipers, and some mountain mahogany burning here, smoldering there, smoking everywhere. But it had potential, and we attacked it with whatever vigor our steep climb had left us.

Once we had a line around the fire, Dick advised the ranger station by radio that we had it contained and controlled. As we began mopping it up, Dick commented on some real hot spots and allowed as how "We sure could use some water on this job before it gets dark."

Knowing *we* meant *me,* I gazed at the pickup far below. "I'll go get it."

"Right," Dick agreed, and added as I started down the mountain, "Bring both of 'em."

Both of them! Water weighs eight pounds a gallon. Ten gallons of water weigh eighty pounds, twice as much as a smokechaser pack. I knew I was in for a long crawl and a half back up the mountain with those two water packs.

It was dark by the time this firefighter had scrambled back to the fire with that ten gallons of water. It was cold by the time the two firefighters had used that water to cool off the last of the hot spots. It was darn cold by the time the two firefighters, who knew better than to try to get off those crags at night, had kindled a small fire inside the fireline to heat their rations, boil water for coffee, and warm their hands. And it was colder still when, at first light, they declared the fire out, shouldered their gear, and made their way off the crags toward the pickup, Bridgeport, and a hot two-dollar breakfast at the Sportsman's Inn as it opened for the day.

All the fires that summer were small. Most were caused by lightning. Scattered around the district as they were, each was a chance to learn more of the country as well as hone fire fighting skills. And some afforded other interesting experiences. On a fire at Masonic Gulch, for example, Dave, a University of Utah student on the trail crew, and I found—and left in place—a yellowed paper tucked inside a rusty Prince Albert tobacco can and placed inside a cairn to document an old mining claim. That find made the time of the old-timers seem not so long ago. And, indeed, it wasn't. Some were still around.

I learned a lot that first summer—and in succeeding summers—from some of those old-timers. Most notable, I guess, was Ella Cain, a true daughter of the Gold Rush born in nearby Bodie in 1882. "Bodie deserved and sustained its reputation of being the most lawless, the wildest and toughest mining camp the Far West has ever known," she wrote of her home town in her 1956 book *The Story of Bodie*. After college, she returned to teach school and, in 1904, married David Victor Cain, son of the James S. Cains, prominent and prosperous Bodie pioneers. Their son ran the Bridgeport General Store, famous for its exhibit of her Indian basket collection. Mrs. Cain was a great source of local history because she'd lived it – in Bodie and in Bridgeport where her family moved in 1888 when her father, M.J. Cody, was elected sheriff of Mono County.

On her ninth birthday, June 9, 1891, young Ella witnessed what she recorded in her 1962 book *The Story of Early Mono County* as "the most tragic and terrifying event...Bridgeport...has ever known." That was the beheading and dismemberment of local Chinese merchant Ah Quong Tai by some of the several hundred Paiutes who had gathered in Bridgeport to see him tried for murdering, dismembering, and feeding a member of their tribe known as Poker Tom to fellow tribesmen. Acquitted for lack of evidence, he was dragged from the courtroom to a field north of town and dispatched. The rest of the town's history was less violent.

Many of the fires in the Mono Basin in those days were reported by Charlie Hendel, an old-timer who with his wife lived northeast of Mono Lake at a place called Mark Twain Camp, just off the Pole Line Road and a couple miles east of the California-Nevada state line. Charlie and his wife had built their house there of surplus ammunition crates from the U.S. Navy ammunition depot near Hawthorne where he worked. A one-time Nevada state assemblyman, he'd been key to getting that road connecting his Hawthorne constituents with the Mono Basin paved. The dedication of the completed project at the state line in 1957 was a big event in his life. Forest Service fire crews were always welcome at Charlie's place, and Maurice, Hank, and I spent a July night there before enjoying a big breakfast at the Hendels' table and walking in to a small lightning fire on Cedar Hill at daybreak. Charlie never tired of spinning yarns about his early years as a cowpuncher and chuck wagon driver.

There were many others. As a young man early in the twentieth century, Hal Eaton, owner and operator of the Redwood Motel, had helped his father Fred Eaton acquire Owens Valley land and water rights for City of Los Angeles water and power czar William Mulholland. Les Phillips, Mono County road crew foreman born in the Bodie Hills mining town of Masonic where his father J.S. Phillips was partner in a large mine, had been a Civilian Conservation Corps crew foreman in Yosemite National Park during the Great Depression. He was Marion Hysell's fishing partner, and full of local lore. And legendary Mono Basin rancher and Bridgeport District grazing permittee Richie Conway was as ridin' and ropin' a ranchman as ever there was. I could have learned much more from these old-timers had I been smart enough to make the effort, but I was young and only beginning to develop an historical perspective.

Most of the summer, of course, was spent on work other than firefighting. Depending on the fire danger, part of the crew usually worked in the field, part at the ranger station. Work in the field was mostly campground cleanup and maintenance, sign erection and maintenance, and other jobs from which we could be recalled or dispatched quickly. Work at the station involved fire equipment maintenance and sign construction in addition to general duties like cutting the grass, cleaning the shops and warehouses, and painting.

Some of this work afforded time to think about life and figure things out. One day, while cleaning campground outhouses, I found myself thinking about the chain of command and where I fit into the grand scheme. President John F. Kennedy, of course, was at the top, and Secretary of Agriculture Orville Freeman was his cabinet officer who ran the Department of Agriculture. Chief of the Forest Service Edward P. Cliff reported to him, Regional Forester Floyd Iverson reported to him, Forest Supervisor Ivan Sack reported to him, Ranger Hoag reported to him, Marion Hysell reported to him, and I worked for Marion. Except for the toilet brush in my hand, I was low man on the totem pole!

Marion Hysell was at the center of this work. True to his word, the FCO worked me—and the rest of the crew—hard. But nobody worked harder than he did. Much more leader than boss, he led by example. His easy yet effective leadership helped us work, enjoy the work, and appreciate what we accomplished. And his infectious sense of humor and unique outlook made it all fun.

I particularly enjoyed the public contact aspect of my part-time district clerk work. Most ranger station visitors genuinely appreciated my assistance. Sometimes that assistance involved more than information, directions, and permits.

Late in the morning on one of my office days, a worried looking visitor drove in and asked if I could help him with his dog.

"What's wrong with your dog?" I asked.

"Porcupine quills," the man responded as he produced a whimpering half-grown beagle with a face full. "Got 'em at the campground a little while ago."

"Marion Hysell was at the center of this work."

Fortunately for me, I wasn't alone at the station that day. Marion Hysell was getting ready to shoe Old Blue.

"Let me get some help," I said, not committing myself further as I walked toward the barn. "I'll be right back."

I knew Marion would know what to do, and he did. We must have made quite a picture, sitting there on the front porch of the little office. As the beagle's master looked on, wringing his hands, Marion skillfully extracted each quill as I held the squirming puppy.

"Thanks a lot!" the man exclaimed as he climbed into his car with the dog. "I knew I'd come to the right place."

The right place. That squared with my idea of what a Forest Service ranger station should be—and should always strive to be—the right place to go for help and get it.

And so the summer wore on. All in all, it added up to a few fires, a lot of work, and not much time off.

I worked every day but three that summer. Two of my days off were early in the summer. On the first, I caught a ride to Mono Village and walked up the Horse Creek Trail into the Hoover Wild Area—which would be designated the Hoover Wilderness with passage of the Wilderness Act in 1964—as far as the snow that still blanketed this alpine wonderland allowed. The second was spent at the bunkhouse, reading, writing, and washing my clothes. My family—Mom, Dad, and brothers Dave and Richard—passed through late in July, and I was granted a third day off to show them the country. It was a different kind of country—even for a well-traveled naval officer's family—and a different kind of life.

"Where's the forest?" asked Mom, who had grown up in the Pacific Northwest, as she gazed across the sere eastern Sierra landscape.

"Up there," I pointed toward the Sierra. "I'll show you." And, on that day off, I did. I wanted Mom and Dad to appreciate this country as I was beginning to appreciate it, or at least to understand my job and why I liked it. They had been skeptical about the job from the outset, and still seemed to have reservations. Dad, who had spent a year in the Civilian Conservation Corps before joining the Navy in 1934, seemed the more comfortable about my situation. At any rate, I was happy to see them, and happy for them to see me there.

As the fire danger increased in late July and remained high through August and into September, the fire crew was kept close to the ranger station even after regular working hours. In fact, we were put on two hours of fire standby duty—after an hour break for dinner—during which we continued to work around the station. A lot of this fire standby time was spent peeling the bark off green lodgepole pine logs, stacking them to season for a year or two, and treating seasoned logs with preservatives. The green logs peeled quickly. The bark on most could be "unzipped" the length of the log with the sharp tip of a fire shovel blade, then flaked off in large sheets in a matter of a few minutes. When necessary, a drawknife was used to finish the job. These logs were used for building signs and other structures. We showered well and slept better after those ten-hour days.

"A lot of fire standby time was spent peeling the bark off green lodgepole pine logs."

After-hours bunkhouse life that summer was a routine of preparing and eating dinner, washing dishes and cleaning our quarters, and doing our laundry. Our only contact with the outside world was Dick's little radio, which ran well enough on generator power. Television had not yet arrived in the Bridgeport country. Occasional evenings in town and congenial "bunkhouse inspection visits" by Ranger Hoag provided variety.

About once every week or two, John's dad, Dr. Jack Pasek of Minden, Nevada, stopped by to drop off venison, chat, and dispense free medical advice. Beer, in moderation, of course, was much better for us—had more food value—than soda pop, he held. His two-state general practice extended at least 125 miles up and down U.S. Highway 395 from the new Mono Medical Center in Bridgeport north into Washoe County, Nevada.

Every evening I honored the Forest Service tradition—this one ended in the early 1960s—of recording my day's work in an official diary. Kept by each employee to document his work, these diaries were given to the district clerk at the end of each two-week pay period. Pay slips, based on these diaries, were sent to the forest supervisor's office in Reno. My bunkhouse mates, somewhat less religious in their observance of this ritual, usually reached the end of a pay period with fourteen blank little yellow pages. So, once every two weeks, we'd sit around the kitchen table and reconstruct the pay period from my diary. "Les, what was it I was doing while you and Dick were on that Devil's Gate Fire?"

The summer of 1962 ended all too soon, and it was time to return to the Bay Area and a second year of college.

I boarded the Greyhound bus in Bridgeport on a cool September evening. Five minutes after pulling out of town, the bus sped past the Bridgeport Ranger Station. It was dusk, and lights burned in some of the windows. My eyes stung, and I quickly wiped away unexpected tears. It had been a more meaningful summer than I could begin to understand then or explain now. I had done something I had always wanted to do. And, without knowing it, I had been assessing that summer and myself. I realized then that I had lived up to some of my expectations of myself, but not to others.

I guess I had lived up to the Forest Service's expectations. Ranger Hoag had asked me to come back the next summer as the district's fire prevention guard. It was a job I couldn't know then I would keep through the 1966 fire season.

Chapter Three: Fire Prevention

The Bridgeport Ranger Station had changed considerably by the time I returned in June, 1963, for my second season of work—and my first as district fire prevention guard.

A new, modern office building, complete with a combination lobby and clerk's office, three other offices, and two rest rooms had replaced the old one-room office building that had been moved to one of the Toiyabe's other districts. Closer to and more visible from U.S. Highway 395, this office, with its soon-to-be-paved entry road and parking lot, attracted many more visitors.

Ranger Bob Hoag and family had moved into a new ranger residence, and Marion Hysell and his growing family had moved from the converted bunkhouse they had occupied for three years into the old ranger residence. Range Conservationist Gordon Decker, in turn, had occupied the Hysell's old quarters. Southern California Edison had run power lines up the highway, so the station no longer depended on a gasoline generator for electricity. There was a new dump truck for project work and hauling campground garbage. And, for fire control, there was a new 200-gallon tank-pump unit mounted—along with fire tools for a whole crew—on a new Chevy truck. This would be my patrol vehicle. In addition to my fire prevention duties, I would be the district's initial attack fireman as long as I was driving that rig. And, speaking of my duties, a new district clerk named Anita had relieved me of that particular job.

The fire prevention guard job—and the promotion to GS-4 that went with it—suited me just fine.

There are two ways to prevent wildfires: risk reduction and hazard reduction. In the official Forest Service fire control lexicon of the day, "risk" was defined as "the chance of a fire getting started as determined by the presence and activity of causative agents," and "hazard" as "a fuel complex defined by the kind, arrangement, volume, condition, and location that forms a special threat of ignition or of suppression difficulty." In plain English, fires could be prevented by reducing ignition incidence and reducing fuel availability and flammability. The situation on the Bridgeport Ranger District—a large district with a staff too small to do much about hazard reduction—dictated emphasis on risk reduction. There was no way to prevent lighting-caused fires, of course, so the fire prevention challenge narrowed to preventing man-caused fires.

Since fires caused by human carelessness are the fires that can be prevented, the main idea of fire prevention is to prevent fires through people. My job, as fire prevention guard, was to reach as many people as possible with an effective fire prevention message. The basic approach to that job on the Bridgeport District included public contact patrols, posting fire prevention and fire regulation signs, and—when necessary—enforcing national forest fire regulations and state fire laws.

The greatest risks of human-caused fires existed where the most people—"causative agents"—were concentrated. The heavily used Twin Lakes recreation area, with several developed national forest campgrounds strung along Robinson Creek as well as three resorts and at least a hundred summer homes on both national forest and private lands, was at the top of this list. Other significant visitor concentrations were in the Virginia Lakes and Green Creek areas of resorts, pack stations, summer homes, and campgrounds, and in undeveloped camping areas like the Virginia Creek, Buckeye Creek, and Swauger Creek canyons.

Following fairly routine routes, I patrolled these higher risk areas as often as possible during the fire seasons of 1963 through 1966. The most frequent patrol involved an early morning or late afternoon visit to Buckeye Canyon on the way to or from the popular Twin Lakes area where the balance of the day was spent. A patrol to

The Bridgeport Ranger District fire patrol truck at Twin Lakes. This pickup with a slip-on tank and pump unit was the district's only fire rig in the mid 1960s.

the Green Lakes and Virginia Lakes areas, usually returning to U.S. Highway 395 at Conway Summit via the old Virginia Creek road, was another day-long job, while a patrol to the Devil's Gate-Swauger Canyon area or down the East Walker River into Nevada—places where there were no resorts or developed campgrounds but where visitors sometimes camped—could be completed in a morning or an afternoon. Of course, when there were signs to install, repair, or re-paint, the patrols were modified. I patrolled the more-remote, less-visited reaches of the district less frequently.

I don't know that any fire prevention guard ever prevented a fire just by riding around in his truck. I know I didn't, because I didn't. The key to these fire prevention patrols was public contact, and the key to public contact was to get out of the patrol truck and talk with people. So, instead of just driving though campgrounds and raising dust, I routinely parked and walked the campgrounds. That made me available to visitors, either to initiate conversations or to respond to their questions. Outside developed campgrounds, I parked a respectful distance from and walked into camps to greet the campers, answer whatever questions they had, and work my fire prevention message and campfire permit check into the conversation. I did the same along heavily-fished streams. Another way I made myself available to the public was by patronizing the resort restaurants along my patrol routes. This cost more than packing a sandwich for lunch, but it was a rare noon hour that I didn't make at least a few good contacts.

Availability's partner in the public contact business is visibility, and I made it a point to look the part, to be easily recognized as a Forest Service representative. A good haircut and a clean shave, a freshly-pressed uniform shirt with badge in place, clean green jeans, and solid work boots, coupled with a pleasant disposition, projected a positive image. There is no place in public service, I agreed with the district ranger and the FCO, for slovenly appearance and bad manners. I've no proof, of course, but I'm convinced this friendly, helpful, face-to-face approach contributed to my success at the job.

Availability and visibility were enhanced by other patrol route chores. Chief among these was sign installation and maintenance. I often left on patrol with new road signs or fire prevention signs to be installed somewhere on the district lashed to the rig. In addition to the tank-pump unit and fire tools, I always carried digging tools for this job as well as paint brushes and a five-gallon can of brown stain to keep all district signs looking fresh. Since signs were almost always installed along roads, I was available and visible and made many valuable public contacts while installing and maintaining them. This approach, I'm convinced, made a positive impression on the public.

And, early on, I learned that, while my job was fire prevention, I was expected to know just about anything anyone might want to know about anything even remotely connected with the Toiyabe National Forest and the surrounding country. All I'd learned during that fire crew summer of 1962 really paid off then.

Among the more frequently asked questions, of course, was "Where's the good fishin'?" At first, that was a hard one for me because I didn't fish. And though I never did, I eventually got to know enough to answer most fishing questions. Sometimes, if I'd figured the asker as a person with a good sense of humor, I'd quote—anonymously, of course—the great fisherman I worked for: "There ain't no good fishin', just good fishermen."

One of the more challenging aspects of my public contact work was explaining the difference between national parks and national forests to visitors. Many asked why adjoining national parks and national forests, like Yosemite National Park and the Toiyabe, were run under different rules by different rangers, and even why Smokey Bear protected national forests from wildfire but dared not roam in national parks. Only a few cared to wrap their vacationing minds around more than my shortest and simplest explanation of the differences between the National Park System and the National Forest System and park rangers and forest rangers.

To those who did, I was able to explain—in as much detail as they cared to hear—that national parks are areas of exceptional scenic, historic, or recreational interest or value set aside by Congress and administered by the National Park Service, a agency of the U.S. Department of the Interior, to provide for public enjoyment "in such manner and by such means as will leave them unimpaired for the enjoyment of future generations." They are, I explained, great outdoor museums in which the only development allowed is that needed for administration and protection or required for the comfort and convenience of visitors. In refining this explanation, I got to the point I could just about quote the official government folder on the purposes and management of national parks and forests verbatim. In national parks, I told them, "the basic policy is to preserve nature...while providing...for visitor appreciation and...use." Thus, in national parks:

> All animal species receive equal protection, subject to sensible controls; virgin forests remain unlogged to go through their natural cycles, grazing is limited and is being steadily decreased, with the ultimate objective of eliminating it completely; lands, with a few exceptions specifically authorized by Congress, are not subject to mineral entry; impoundment of lakes or streams for irrigation, hydroelectric power, or other purposes is opposed.... The basic policy is to preserve nature as created while providing areas for visitor appreciation and intelligent use.

National forests, I would continue, are administered by the Forest Service, the largest U.S. Department of Agriculture agency, under the principles of multiple use and sustained yield, also defined by the folder.

> Multiple Use means that resource management is coordinated so that areas of land produce a combination of values that best serves the American people. Sustained Yield means that resources are managed so as to provide services and products at a level of supply as high as can be sustained without harming the land's ability to produce.

By the time I finished, those who'd stuck with me understood something about the benefits of managing national forests under those principles. These benefits also were described in the folder.

> Water from the forests irrigates millions of agricultural acres and is used for domestic and industrial purposes in hundreds of cities. Timber from the forests is manufactured into 5,000 products sold worldwide. National Forests and Grasslands provide habitat for millions of big and small game, game birds, and fish, and produce forage for 1,300,000 cattle and 2,600,000 sheep. Forest recreation includes scenic drives, wilderness travel, picnicking, camping, hiking, skiing, swimming, boating and, subject to State conservation laws, some of the country's finest hunting and fishing.

That was one handy folder! I explained all this as clearly and briefly as possible, and without resorting to a lot of jargon and statistics.

Only rarely did visitors raise more esoteric questions about various resource management issues or ask why the Forest Service is in the Department of Agriculture while the National Park Service and most other federal land management agencies are in the Department of the Interior. When they did, I usually had the answers.

Most folks, however, wanted basic answers. They were visiting the Toiyabe National Forest for recreation, not education; for pleasure, not politics. I knew my job was to answer their questions, not to inflict unwelcome quantities of information. I let them set the pace.

Since much of the Bridgeport District—even at higher elevations—was rangeland, many of the questions I fielded dealt with livestock grazing on national forests. One afternoon in Buckeye Canyon, for example, I met a family of four backpackers returning from a several-day trek into the high country. They'd had a great time, the prosperous-looking head of the family told me, but he wanted to talk about cattle. The round tin cup hanging from each family member's belt gave me more than a hint of his opinion on the subject.

"Why are these cattle here?" the man demanded to know, waving his arm toward the Big Meadows.

"This is a national forest grazing allotment," I answered. "The rancher who owns those cattle holds grazing privileges and pays a fee to graze them here."

"They're ruining the country," he countered. "They shouldn't be here. Why are they allowed to be here?"

I didn't get far with my explanation of the multiple-use concept under which our district range conservationist labored to manage cattle and sheep allotments in a way that would maintain the quality of the forage resource as well as the integrity of the watershed, wildlife habitat, and recreation resources. He didn't care that some ranchers depended on national forest range to stay in business. He just wanted those cattle out of there, pronto. And, although there were no active timber sales on the district, he wanted national forest timber cutting stopped, too. Basically, he wanted national forests to be national parks.

So I took a different tack. After assuring him I shared his concerns for proper public land management, I used the next few minutes of our conversation to find out more about him. I soon found out that this gent was a corporate executive who lived the good life in Palo Alto. He and his family resided in a rambling, contemporary redwood home and dined on juicy steaks at fashionable restaurants college students certainly couldn't afford.

"You know," I suggested, "the resources that make the way you and millions of other American live possible have to come from somewhere."

"Look, I see what you're driving at," he responded. "But they don't have to come from places that should be wilderness areas."

"Places like the redwood coast? Places like public range?"

"Since much of the Bridgeport District...was rangeland, many of the questions...were about livestock grazing on national forest land." A cattle allotment in Buckeye Canyon, and a sheep trailing along Green Creek.

"Exactly...." Just for an instant, his eyes showed he had made the connection, he had recognized the conflict between his affluent lifestyle and his environmental convictions. But only for an instant, because he wasn't ready to accept—or, at least, concede to me—the fact he couldn't have it both ways forever. "The Forest Service should get those cows out of here!" were his last words to me. But he sounded more thoughtful, less arrogant, less certain.

I was less certain, too. And, as the family walked away and I drove away, I began to think anew about the split in the American conservation movement between John Muir's preservationists and Gifford Pinchot's utilitarians that was beginning to define the debate of the coming environmental revolution.

Sheep grazed many of the district's range allotments, and I got acquainted with several Basque sheepherders. Immigrants from the Pyrenees region of France and Spain, Basques had herded sheep for large sheep outfits for several generations. Some came to the United States to work for a few years, save their wages, and return to Europe to marry. Others stayed, found other occupations, and became citizens. Most sheepherders camped on the range in pyramid-shaped tents, and many still moved camp gear on pack animals. Their culture was new and strange to me, and I was surprised at and wondered about the apparent antipathy some Basques from north or south of the border that split their homeland seemed to have for their ethnic brothers. "Goddamn Frenchy Basco!" one once called a herder of the opposite nationality while talking with me. Maybe he was just kidding. I still don't know.

There were, of course, many other jobs associated with my patrols. One of these was annual special use permit and fire safety inspections of the sixty-five or so summer homes located on national forest lands at Twin Lakes, Green Creek, and Virginia Lakes. It didn't take me long to learn enough about structural fire safety to help the fine folks who owned these summer places eliminate some dangerous conditions. Storage of flammable liquids in glass containers, it

"There were…many other jobs associated with my patrols. One of these was annual special use compliance and fire safety inspection visits of…summer homes on national forest lands."

"Another routine patrol job was stocking and maintaining fire tool caches... "such as this one being inspected by Fire Control Officer Marion Hysell.

seems, was the most frequent indoor fire safety problem. The leading concerns outside included flammable grass and brush—flash fuels—too close to the structures and, as the lessons of the 1961 Bel Air-Brentwood conflagration in Los Angeles began to sink in, wood shake shingles.

Another routine patrol job was stocking and maintaining the fire tool caches strategically located around the district—at places like the Hunewill Ranch, the Mono Village store, and The Top restaurant at Conway Summit—for use by Forest Service firefighters and others in emergencies. This was a fire presuppression rather than a fire prevention task, one of many "activities in advance of fire occurrence to insure effective suppression action" as the official definition of the term read. Presuppression activities, especially maintaining fire suppression equipment, were also one of my big jobs whether on patrol or at the ranger station.

Communications, essential to safe and successful patrols and all other field work, came in two forms, the Handie-Talkie—a portable radio in a padded satchel about the size of a briefcase, it had a detachable microphone and an extendable antenna—and the dashboard-mounted vehicle radios. Since I was likely to be away from my rig at any time, I always had a portable to pack with me as well as to back up the truck radio. These were more or less "line-of-sight" radios, and gaining a little elevation could make all the difference when on a trail or road up some canyon. We often scrambled uphill to be able to transmit and receive, and more than once I was seen standing on the cab of my patrol truck, mike in my right hand and holding the Handie-Talkie above my head with my left to "get through" to the ranger station. When in the field, we checked in hourly with a position and status report. Standard "four-codes" and "ten-codes" were used to keep transmissions brief.

On more than one occasion while I was on patrol, the ranger station radioed instructions to locate visitors in one of the campgrounds and inform them of a family emergency. Once, in town after work, I spotted a vehicle I'd been looking for all day, tracked down a member of the party, and delivered the message. And, when riots erupted in the Watts section of Los Angeles during August 1965, I was sent to locate Los Angeles city police officers and county sheriff deputies camping on the district who were being recalled from their vacations for the emergency.

Anything could happen on patrol, but the basic job was to prevent wildfires.

My greatest fire prevention ally was Smokey Bear. That's backward, of course. I was Smokey's ally, and only one of many. At any rate, Smokey had been America's "forest fire preventin' bear" for nineteen years by the time I had become the Bridgeport District fire prevention guard in 1963, and with his famous "Only you can prevent forest fires!" tag line had become the most-recognized symbol in advertising history.

I was proud to be on Smokey's team, and used the Cooperative Forest Fire Prevention Program materials I ordered each year to the fullest. I soon—albeit secretly—fancied myself "the fastest staple gun in the West" for the way I kept fresh Smokey Bear posters displayed on well-maintained peak-roofed signboards—all strategically located, by the way, in accordance with a fire prevention sign plan—all over the district. Posters were also displayed in Bridgeport's public buildings, stores, restaurants, motels, and gas stations, and in re-

"I soon...fancied myself 'the fastest staple gun in the West' for the way I kept fresh Smokey Bear posters displayed...all over the district."

sorts and businesses throughout the district. A supply of Smokey Bear comic books for children met while on patrol was always in the patrol truck along with maps and other information for public distribution, and every campfire permit issued on the district had a small Smokey Bear fire prevention reminder stapled to it.

I continued the district's practice of having Smokey Bear appear in Bridgeport's famous Fourth of July parade, usually cajoling a member of the fire crew or trail crew to wear the costume ordered from the regional office in Ogden. After the parade, a properly-escorted Smokey always mingled with the crowd in town and visited Robinson Creek campgrounds. During the summer of 1964, I ran a periodic Smokey Bear Junior Forest Ranger School for resort and campground children.

Still another way I reached the public was by writing news and feature articles for the *Bridgeport Chronicle-Union* and the other eastern Sierra weeklies published by the Chalfant Press in Bishop. All these articles—whether warnings of current fire danger, reports of recent fires, or features about some aspect of fire control or even about national for-

"I continued the district's practice of having Smokey Bear appear in Bridgeport's famous Fourth of July parade... and at Twin Lakes area campgrounds."

Fire Burns 350 Acres Along Robinson Creek

BRIDGEPORT — Human carelessness ignited the Robinson Creek Forest Fire which blackened about 350 acres of private and Toiyabe National Forest timberlands near Bridgeport June 29 and 30. According to Les Joslin, Bridgeport District prevention guard.

A special project fire overhead team from the Forest Supervisor's office in Reno arrived and a fire camp, the nerve-center of a large fire organization, was established. A mobile fire weather unit provided up-to-the minute reports on fire weather conditions.

IT COULD HAPPEN AGAIN

National Forest Officials Warn of Coming Danger

One year ago this week, on June 29 and 30, 1964, the Robinson Creek forest fire blackened about 350 acres of private and Toiyabe National Forest watershed and timberlands near Bridgeport.

many decades will pass before this scar is erased from the landscape.

Probably the most tragic aspect of the Robinson Creek forest fire

★ **BARNEY LAKE—is but one of many beautiful mountain lakes in the Hoover Wild Area of the Toiyabe National Forest which affords outstanding recreational opportunities.**
(Photo by Bob Hoag).

Hoover Wild Area Offers August Recreation Here

by Les Joslin
U.S. Forest Service

The month of August, in which the summer vacation season is at its peak, attracts many visitors to the already crowded campgrounds on the Bridgeport District of the Toiyabe National Forest. In order

people. Here, amidst majestic Sierra peaks and towering pines, one may find spiritual solitude and a refuge from civilization available only in the wilderness.

You don't have to be a hardy hiker or mighty mountain man to take advantage of this Sierra wonderland. Much of it is easily accessible on trails built and

First Man-Caused Fire Reported In Bridgeport Area

The first man-caused forest fire of the 1963 summer fire season on the Bridgeport District of the Toiyabe National Forest occurred Monday, July 22.

Labor Day Visits Increase Forest Fire Danger

BRIDGEPORT — The tremendous influx of forest visitors and recreationists to the Bridgeport District of the Toiyabe National Forest over the 1964 Labor Day weekend, the last big holiday of the summer season, will greatly increase the risk of man-caused

Fire Danger Mounts In Bridgeport Area

A dangerous combination of cured cheat grass, other flash fuels, and increased recreation use on the Bridgeport district of the Toiyabe National Forest have

Lightning Fires Hit Toiyabe Forest

BRIDGEPORT — Five lightning fires caused which burned a total of about 22 acres of pinon and juniper-covered watershed lands, were controlled by U.S. Forest

Fire Danger Emphasized

"The danger of man-caused forest fires is normally greater in the month of August than at any other time during the year," stated H. Marion Hysell, Forest Service fire control officer on the Bridgeport District, Toiyabe National Forest.

Warning Issued On High Fire Danger In Back Country

Forest fire danger in the High Sierra is becoming extreme as the summer progresses and the high

★ **AERIAL FIRE PATROL—H. Marion Hysell, fire controll officer on the Bridgeport District of the Toiyabe National Forest, and pilot Bob Whittenberg review their flight route before leaving the airport in Bridgeport on an aerial fire patrol.**
(Photo by Lloyd Parratt).

Fire Patrols Made During Storms by Forest Service

In its constant vigil to protect the Nation's forests, ranges, and wild lands from wildfire, the Forest Service of the U.S. Department of Agriculture uses many methods of fire detection, one of which is aerial patrol flights.

— a tough job at best. By being careful with fire in the forest, you can help the Forest Service in its campaign against wildfire in the forest and on the range. Lightning

Fire on Bald Mountain Burns 350 Acres

A delayed-action lightning strike ignited the Wichman Forest Fire which blackened about 350 acres of beetle-killed pinion pine on the southwest slope of Bald Mountain on the Toiyabe National Forest, according to H. Marion Hysell, Forest Service fire control officer on the Bridgeport Ranger District.

Bridgeport Area Fire Danger Is High, Ranger Says

Forest fire danger on the Bridgeport District of the Toiyabe National Forest and adjacent areas remains extreme as hot weather and dry winds continue to draw the moisture from the countryside. These extreme fire danger conditions favor ignition

Man Caused Fires Plague Bridgeport Area

BRIDGEPORT — Four man-caused forest fires were suppressed on the Bridgeport District of the Toiyabe National Forest during the last two weeks, bringing the District's total number of man-caused fires this summer to

"Still another way I reached the public was by writing news and feature articles for the Bridgeport Chronicle-Union *and the other eastern Sierra weeklies...."*

est recreation opportunities—pushed fire prevention. I also slipped fire prevention pleas into the evening visitor information programs I put on from time to time at the Crags Resort, Doc and Al's Resort, and the Hunewill Guest Ranch. It was practically impossible for anyone to live near or visit the Bridgeport District without becoming aware of Smokey's forest fire prevention message.

I was, of course, motivated by the FCO's example as well as by my own enthusiasm. Marion Hysell received a Smokey Bear award in 1965 for the large log fire prevention signs featuring Smokey he had designed and, with the help of the fire crew, constructed and erected along U.S. Highway 395 and the Twin Lakes highway.

I spent a few days each summer in the Hoover Wilderness, where increasing back-country travel was compounding the risk of fire. Otherwise, wilderness patrol and public contact were left to the district's three-man trail crew.

Most of my wilderness patrols were on foot, but on rare occasions I rode a horse. Now, I never claimed to be a cowboy. My only previous experience with horses had been gained during a couple summer visits to an uncle's farm, and most of what I knew about them had been learned from books and movies. So there I was again, long on theory and short on practice. But I knew enough to get along and, in spite of my shortcomings as a buckaroo, most of those horseback patrols went well.

In 1965, for example, I enjoyed a patrol of the rugged and scenic Green Creek drainage with botanist Lloyd Parratt as he conducted plant surveys. Lloyd, a Chaffey College professor with seventeen seasons as a National Park Service seasonal ranger-naturalist at Glacier National Park and five at Yosemite National Park before that, was now working summers for the Forest Service and was my roommate for the summer. While he botanized, I talked with occasional wilderness travelers and learned the lay of the land between Green Lake and the Yosemite National Park boundary at Summit Lake. I previously had known this beautiful corner of the district only from maps and patrol plane flights. A mountain-wise black mare, appropriately named Coaly, afforded good transportation and no trouble on this trip.

"I was…motivated by the Fire Control Officer's example…. Marion Hysell received a Smokey Bear award in 1965 for the large log fire prevention signs he had designed…, constructed and erected…."

"Most of these horseback patrols went well."

I say "most of these horseback patrols went well" because I darn near wound up finishing one of them on foot.

That was the time Marion Hysell and I rode up to a little unnamed lake above Barney Lake and below Hunewill Peak to rebuild rock fireplaces and erect a sign. After finishing the work and eating lunch, we began the return trip on the two big black horses hired from the Mono Village pack station for the day's work. About half way down to the junction with the Barney Lake trail, Marion and I met a group of hikers. They were bound for the little lake we had just left, and we dismounted to talk with them. About what, I can't remember. What I can remember is that my mount suddenly decided to quit the country. He reared, yanked the loosely-held reins from my hand, and launched himself down the trail toward the pack station more than four miles away.

"Now you're a hiker, too!" one of the hikers observed with thinly veiled satisfaction.

Instead of explaining my preference for shank's mare, I excused myself with something like "Darned if I'll walk back to Mono Village!" and lit out after the horse.

Crown Point, at 11,346 feet, looms above Barney Lake in the Hoover Wilderness.

"He'll never catch it," another of the hikers predicted.

But there was a chance. I recalled the series of switchbacks in the trail up this steep slope, and hoped they'd work in my favor. I plunged off the trail and down slope over boulders and through thickets to head off the fugitive mount.

The first time I tumbled back onto the trail I was just too late. Down the trail to my right, the horse was hightailing it into the next switchback. Back into the woods I plunged, again careering downhill as aspen branches slapped my face and mountain mahogany slowed my progress.

Within moments I was back on the trail, and this time ahead of the game. Off to my right, the horse had just turned out of the switchback and was thundering toward me, eyes wild and ears back in what looked a lot like determination. I was determined, too, and as the big black tried to evade me I grabbed its reins just below the bit, yanked down hard, and wrestled it to a kicking, snorting, dusty stop. Its reins tight in my grasp, the horse followed me back up the trail. Would catching this runaway redeem me in Marion's eyes? In the eyes of the hikers?

By the time I had led the recaptured mount back up the trail, where Marion and the hikers were waiting to see if I would walk or ride home, I had managed to brush off dust and leaves and tuck in my shirt. I had also reviewed and rejected every alibi east of the Sierra. The truth of what had happened was obvious. Marion grinned. He seemed satisfied at the fact I had caught the horse. The hikers seemed impressed by the same fact.

We mounted up to resume the ride to the pack station. I told myself I was finished with horses.

But those horses hadn't finished with us. Next was the FCO's turn. Almost back to the pack station, we met a woman and two little girls sitting on boulders beside the trail, and stopped to speak with them. Marion was soon engaged in polite conversation with the woman while the little girls admired our mounts.

"I like horsies," the younger girl beamed at me.

Whoosh! Her words of fondness were drowned as Marion's big black, standing spraddle-legged directly in front of the three, proceeded to relieve himself.

This didn't do much for the conversation, and horses just don't move on while they're responding to nature's call. So there we were, all trying our darnedest not to appear to notice the steady stream splashing and steaming in the trail dust not six feet from the woman and the little girls. Except, of course, the little girls, whose eyes widened at the spectacle. I had to bite my tongue to keep from laughing as I watched Marion, close to laughter himself, and the woman affect an immediate interest in virtually every aspect of our surroundings—Ah! the glories of nature!—save the equine cascade that held the little girls spellbound. Finally, after what seemed eons, the downpour abated. And the big black snorted satisfaction.

"Well, we better be gettin' on down the trail," Marion rather disingenuously excused us, and we rode on. Once around the bend and out of sight and—I'm sure—earshot, we doubled over in our saddles with laughter.

No job is perfect, and the fire prevention guard job was no exception. Burning off the Forest Service dumps, to which campground

"Burning off Forest Service dumps...was a hazard reduction part of the job."

and permittee refuse was hauled, was a hazard reduction part of the job and just about the only part I didn't really enjoy. Why not? Well, it started cold and dark, got dirty and smelly, and finished hot and dry. It involved getting up long before sunup, loading a couple five-gallon cans of diesel fuel onto the patrol truck, driving to the dump to be burned that morning, pushing the garbage and trash piled up around the sides into the pit, anointing the whole mess with diesel fuel, touching it off with a fusee, patrolling the surrounding area for spot fires while the dump burned, and making sure the whole thing was out cold by about ten o'clock before the winds came up. Oh, dump burning had its compensations, like communing with Mono Lake's seagulls who winged north to forage for tidbit supplements to their brine shrimp diets. Once, at the Virginia Lakes dump, I saw a bear. And once, at the Twin Lakes dump, the dump burning job led to an amusing encounter with some fishermen.

Extinguishing the dump fire had drained the pumper tank, so I drove to the nearby Timber Harvest Road bridge across Robinson Creek for a refill. Crossing the bridge, I pulled off the road and backed down to the creek. Not long after I had dropped the draft hose into the stream and reversed the pump, two vehicles halted on the bridge. One was a flashy convertible, the other a battered Jeep. The occupants gestured in my direction, pulled off the road, and parked next to the patrol truck.

At first, I thought something might be up. Perhaps they wanted to report a fire! But the rapid unloading of fishing gear soon betrayed their misguided urgency. This had happened before.

"Hey, are they big ones?" a couple blurted in unison as they scrambled past me to peer into the stream where my drafting hose lay. The rest of the pack followed, clutching fishing rods and tackle boxes.

Feigning ignorance, I responded to their question with another. "Big whats?"

"Rainbows! Man, you know! Trout! Fish! You are the fish man, aren't you?"

"No, I'm sorry, I'm not the fish man," I answered almost apologetically. "Fish are planted by the California Department of Fish and Game. I work for the U.S. Forest Service." Then, after a pause, I added with a smile and only slightly exaggerated pride, "This is a fire truck."

"A fire truck?" They looked unimpressed, to say the least. More like disdainful. But the awful truth was beginning to sink in. "You mean...you're *not* the fish man?"

"Sorry," I shrugged. They probably had been looking for a "fish man" with a "fish truck" all morning. "I'm just filling the tank on this fire truck." I slapped the hose reel affectionately. Then I informed them of the days on which the Department of Fish and Game usually planted fish in Robinson Creek. Needless to say, those fishermen—a dismayed *"that's a fire truck"* look still on their faces—didn't hang around very long.

Afterward, I wondered if I should have told them what I knew of the fish planting schedule. More than a few times, while on patrol, I had found the Department of Fish and Game's large blue and silver "fish truck" off the road in an aspen grove, hidden as the "fish man" sought refuge during his lunch break from the crowds that often chased his truck.

"Why don't you just pipe them from the truck into their frying pans?" I once baited a "fish man" whom I had joined for lunch.

A not-so-routine fire prevention concern came with a research project the Department of Defense and the Riverside Forest Fire Laboratory of the Forest Service's Pacific Southwest Forest and Range Experiment Station were conducting near Montgomery Pass, just over the line in Nevada at the southeastern end of the district. To determine characteristics of "firestorms" that could result from nuclear weapon detonations, researchers had built fifteen "cities"—at least, the fuel concentration equivalents thereof—made up of rows upon rows of immense piles of dry sagebrush, juniper, and pinyon pine that simulated the fuels in various sizes of urban areas. The smallest of these, I was told, contained about ten times the fuel in a town the size of Bridgeport, or in a small city of about 3,500 people. Each of these "cities" would be instrumented to record the desired information when burned. One had been burned in the fall of 1963, after I'd returned to college, and Marion Hysell and a few other district personnel had helped monitor it and patrol for spot fires. Another was scheduled to be burned in 1964.

I saw these "cities" from the patrol plane and then on the ground in July 1964. That really scared me! I was darned if I could figure out how to prevent an unscheduled fire—a wildfire—in these massive piles of fuels in so remote a reach of the district some ninety miles from the ranger station. In addition to lightning, there was a range of human ignition sources. Lightning ignitions, of course, were beyond prevention, and in this case, I decided, human ignitions were, too. As it happened, these fuels were never ignited—according to plan or otherwise—on my watch, and I never did learn the results of this research project.

Still another challenge of the fire prevention guard job had nothing at all to do with wildfire prevention. It had to do with the fact I was sometimes the only public servant around to handle an emergency. It had to do with that sometimes gray area between official duty and moral obligation.

"A not-so-routine fire prevention concern came with a research project... near Montgomery Pass...at the southeastern end of the district. ... One [simulated city] had been burned in the fall of 1963. ... Another was scheduled to be burned in 1964."

Once, on my way to Twin Lakes, I detoured to check the dump I'd burned a few days before. Suddenly, I heard shots, just as the Lone Ranger and Tonto once did at the beginning of almost every episode, and what I saw as I neared the dump scared me. A big, beefy, fortyish man, standing next to a late-model Cadillac sedan, was firing a high-powered rifle—not at me, but at a row of cans and bottles on an earthen berm at the opposite end of the dump. He didn't, I quickly decided, know what was beyond the low ridge toward which he was firing, over which his bullets could fly. And he apparently didn't know that Mono County didn't allow shooting in the area. Somebody had to tell him before it might be too late.

He'd heard me coming, and turned as I stopped the patrol truck. He didn't look particularly threatening. But there were serious unknowns. I didn't know him. I didn't know what he might shoot at. I didn't know he wouldn't shoot me.

"Good morning, sir," I managed as I climbed out of the truck.

He nodded and smiled.

"What are you shooting?" followed, though his targets were obvious.

"Just plinkin' at some cans and bottles."

"Sir, do you know what's over that hill about a mile?" I asked, pointing toward the southeast—the same direction he'd been shooting.

"Just sagebrush, I guess. That's all I see from here."

"No sir. The Hunewill Guest Ranch is over there. I'd guess bullets from that rifle could reach there from here."

"Too many dudes around anyway, right?" he smirked, making light of my concern.

Deciding another appeal to common sense wouldn't hack it, I invoked the law. There was, I told him, a Mono County ordinance that prohibited shooting within the Twin Lakes basin.

"Oh, yeah? Well, we wouldn't want to do anything illegal, would we. Guess we'll just go shoot somewhere else."

And with that he got into his Cadillac and drove away. I was relieved to see him gone. A few minutes later, once my heart stopped pounding, I radioed a report of the incident to the ranger station. He must have left the country. I never saw him again, never heard anything about him. But, most important, nobody had been hurt.

"As I patrolled, I got to know special places...and stories...."
A few mature western white pines stood near Sinnamon Meadow, and a hydroelectric plant at Dynamo Pond powered the stamp mill at the Standard Mine in Bodie.

I didn't know it then, but that wouldn't be my last encounter with an armed visitor.

As I patrolled, I got to know special places in and special stories of the Bridgeport country along my routes. Every corner of the district seemed to have them.

Dynamo Pond, for example, on Green Creek three miles south off U.S. Highway 395, was named for the hydroelectric plant built there in 1892 to power the stamp mill at the Standard Mine in Bodie. Originally, the mill was powered by steam, but by the 1890s wood had become scarce and costly. "Electric power had been developed," Ella Cain had written in 1956, "but was used only at the source." Mine superintendent Tom Leggett "was of the opinion that electricity could be transmitted over wires from a distance," and convinced J.S. Cain and others to invest. Many, including mine stockholders, protested the project as risky, even ruinous. Nevertheless, new electrical equipment was installed at the Standard Mine mill and hooked to the Dynamo Pond power plant. Fearing the power running through the lines would not be able to turn corners and "jump off into space," the engineers ran the thirteen miles of wire from Dynamo Pond to Bodie in a straight line! Leggett made history and his reputation the November day the switches were thrown. "*Yes*, electric power *could* be transmitted for a distance over wires." He and other Bodie project engineers went on to build hydroelectric power plants around the world.

Not far from Dynamo Pond, at a bend in the road overlooking Sinnamon Meadow, stood—and, I hope, still stand—a few, isolated, large western white pines. I couldn't explain then, and cannot now, their presence in that land of Jeffrey, lodgepole, and pinyon pine. I think they were too old to have been planted there by anyone. It's possible, I suppose, they were remnants of a stand established during wetter centuries of the so-called Little Ice Age that began a few hundred years ago, but that's pure conjecture on my part.

Every once in a while a fire lookout or guard reports dust from trailing sheep or cattle as a fire. I almost did worse, and now the story can be told.

I was driving across Bridgeport Valley one night, returning to the ranger station from Twin Lakes and a rare date with a young lady, when an orange glow behind the Bodie Hills gripped my attention. As the glow grew brighter, my foot grew heavier on my Volkswagen's accelerator. By the time I reached Bridgeport and turned north on U.S. 395 toward the ranger station—I could drive the four miles and raise the alarm faster than I could stop at a pay telephone and call it in—I knew we had a big one on our hands.

A quick glance over my shoulder as I turned into the ranger station confirmed that it was a big one.

It was the moon.

Les Joslin and Fire Boss Maurice Crawford on the Stockade Flat Fire in July 1965.

Chapter Four: Fire Suppression

Fire prevention is a funny business. You never really know how good you are at it because you never really know if you've actually prevented a wildfire. You can run a good program—do all the things Madison Avenue advertising experts and good sense say about getting the public's attention and repeating the message—but there's no objective measure of success. Failure is much easier to prove. When a fire prevention guard spots a smoke, it often means he's failed to prevent a human-caused wildfire. And when that's the case, it definitely means he's temporarily out of the fire prevention business and in the fire suppression business.

It had been a snowy winter and a wet spring. In terms of fire danger, that was good and bad. Fuels were moist, but the wet spring of 1963 had produced a bumper crop of flash fuels—mostly that inadvertent Siberian import, cheatgrass—that would cure by midsummer. Ranger Bob Hoag expected a busy fire season, and Marion Hysell's small fire organization was as ready as it could be. I'd been lined out with a full bag of fire prevention duties for my first summer as fire prevention guard.

When the telephone sounded the Bridgeport Ranger Station's three rings about dinner time on July 11, the fire crew figured it was the first fire of the season. Since there hadn't been any lightning, I figured it was the first human-caused wildfire of my three-week-old fire prevention guard career. We were wrong. Marion dispatched me and his fire crew not to a fire but to a flood...sort of...at the Honeymoon Flat Campground on Robinson Creek.

"When a fire prevention guard spots a smoke...he's temporarily out of the fire prevention business and in the fire suppression business."

The crew and I arrived to find a large Jeffrey pine had fallen into the creek and was diverting water into lower-lying campsites. Several camps had been displaced. That tree had to go, and we soon figured the best way to make it go was to put a chain around it and winch it out of the channel. Perhaps anticipating such a need, John had driven his own vehicle, an International Harvester Scout equipped with a winch, to the scene while the rest of us went in a Forest Service pickup.

We had the winch, we had the chain. All we needed was somebody stupid enough to brave the swift, icy waters and hook the chain around the tree. For some reason, the guy who a year earlier had told the district ranger he could type told the crew he was a pretty good swimmer. Some people never learn! I took off my boots, wrapped a safety line around my waist, attached another line to the chain, waded in upstream of the target, and let the current sweep me about a dozen yards to the offending tree. Once there, I pulled the chain to me, secured it around the trunk, and signaled the crew to pull me back.

That they did and, while I shivered in a blanket next to a blazing campfire finding out how hard it was to drink generous campers' hot coffee while my teeth chattered, John revved up his winch and, along with crew muscle, finished the job.

Those campers thought the Forest Service was okay.

Bright and early a few days later, as I was leaving the ranger station office for a day on patrol, I met Forest Supervisor Sack and another distinguished-looking gent in a Forest Service uniform at the front door. Ranger Hoag introduced me to Regional Forester Floyd Iverson who was paying a one-day inspection visit. He looked me in the eye, smiled, and put out his hand. I shifted the Handie-Talkie satchel I was carrying into my left hand, shook his hand, and I left with the impression the Intermountain Region of the National Forest System was in good hands.

On July 22, with not a fire on the district yet, the fire prevention job of the day was hazard reduction. Don, a new firefighter, and I were clearing cheatgrass along the Buckeye Road to remove flash fuels that could carry fire into the canyons. At about half past noon, and halfway through our lunch break, a carload of fishermen, speeding along the road toward the ranger station, honked to a stop to report "a big fire running up the mountain near Eagle Creek!"

Don and I were in the rig and on the way in seconds. As I radioed the four-eleven—emergency fire call—to the ranger station, it occurred to me that this fire was man-caused. There hadn't been a lightning storm in weeks.

We arrived at the fire about the same time the fishermen reached the ranger station with their exaggerated report. I sized up the small blaze burning in brush under a few Jeffrey pines near the Buckeye Pack Station and cranked up the pumper. Then, as Don attacked the flames with water, I hit the radio. The ranger station, on the strength of the fishermens' report, had just ordered an air tanker. I advised the station we would soon have the fire under control and to cancel the air show. The fire was knocked down in minutes. Marion drove out to check up on us, and helped mop up.

The flash fuels cured as expected, and the district's fire danger was high most of the rest of the season. But, after many August lightning storms, that small man-caused fire somehow remained the first and last fire of my 1963 season.

The district wasn't always that lucky.

I arrived at the Bridgeport Ranger Station for the 1964 fire season to learn I was to give a half-hour talk on "Making a Fire Prevention Program Work" a couple weeks later at the first and only Toiyabe National Forest fire school I ever attended. A three-day affair on the Alpine Ranger District, the school began with a morning-long plenary session during which I gave my pitch. The balance of the school, organized like a one-sector project fire on which Maurice Crawford and I were assigned as two of three crew bosses, included instruction in the usual fire control subjects followed by a practice fire.

That training paid off soon. A week later, on June 25, a five-acre fire along Buckeye Creek—apparently caused by a fisherman I'd failed to contact with an effective fire prevention message—proved prelude for the test soon to come.

I was returning to the ranger station from an early morning patrol to Green Creek and the Virginia Lakes when I detected what became the largest wildfire the Bridgeport District would see in my five seasons there.

It was June 29 and, having just passed through Bridgeport, I was northbound on U.S. Highway 395 when, across Bridgeport Valley to the northeast, I first saw it. Was it sheep trailing on Timber Harvest Road? Or smoke? *Smoke!* Wispy, white smoke!

Three miles to the ranger station, and maybe eight or ten to the smoke. I tried to raise the station on the radio as I pushed the gas pedal to the floor. The truck responded, but the station didn't. A glance at my watch told me why. It was lunch time, and with the fire danger rated moderate, the office wasn't manned. This was years before the FCO had a Forest Service radio crackling in his house day and night.

Two miles to the ranger station. I could have switched to channel two, raised the Toiyabe National Forest dispatcher in Reno, and asked him to telephone the ranger station. But I decided I could get there, raise the alarm myself, pick up some help, and be on the way to the still-wispy-but-thickening column of smoke before I'd half finish telling the dispatcher my story.

Two minutes later, I came to a dusty halt in front of the FCO's house and interrupted his lunch in the way firemen hate most. Then, with Don, the only crewman not out on project work, I headed up the Buckeye Road toward the fire. After alerting Ranger Hoag, who decided to man the office and handle communications with the Reno dispatcher until the district clerk returned from lunch, Marion Hysell loaded his Jeep with fire tools. As he wheeled toward the fire via Bridgeport and the Twin Lakes highway, he confirmed my sighting to Ranger Hoag over the radio. Ranger Hoag alerted the supervisor's office and the dispatcher, and ordered air tankers.

Fifteen minutes after leaving the station—probably twenty or so after I first saw the smoke—Don and I were speeding down Timber Harvest Road toward a blaze that, at about a mile's distance, I estimated at ten to fifteen acres and spreading fast through cheatgrass and sagebrush toward the Jeffrey pine-clad slopes of Sawmill Ridge. Through the smoke, on the far side of the fire, a few people could be seen making vain attempts to stop it.

In what was a futile—and probably foolhardy—attempt to head off the fire before it could reach the timber, I left the road and, within a minute or two, Don and I were attacking flames running upslope along a two-hundred-yard to three-hundred-yard front. But two guys with a couple hundred gallons of water and hand tools, it rapidly became obvious, were no match for a fast moving fire of that size.

About the time I recognized that truth, and also realized that my poor headwork was getting Don and me into more than a little trouble, I saw the FCO's Jeep jounce over the sagebrush and halt some distance off. "Fall back! Let's get this rig outta here!" he ordered as he ran our way. So, as I reeled in the hose, Marion maneuvered the patrol truck toward safety. Within a few minutes, the two vehicles and the three of us were back on Timber Harvest Road, and the scene of our "initial attack" was engulfed in flames.

The Robinson Creek Fire, as seen from Bridgeport on the afternoon of June 29, 1964, burned 350 acres on Sawmill Ridge.

Ranger Hoag's pickup crossed the bridge over Robinson Creek, and he joined us on the road as the fire, now all of an hour old, raced into the timber. Taking charge, he requested a spot weather forecast from Reno and, with the FCO, planned for the reinforcements starting our way.

Since little, if anything, could be done about the head of the fire—then moving rapidly upslope, torching and spotting in the timber—two rag-tag crews of mostly pick-up firefighters were deployed along its flanks to prevent it from spotting and spreading to the nearby resort and campground. A local resort owner had begun bulldozing a fireline around the fire's lower end. From my smoke-choked vantage point with the Timber Harvest Road crew, the fire seemed to be living up to the expectations of the spot weather forecast. Under prevailing conditions—a temperature of 77 degrees, a relative humidity of 15 percent, a 15 mile-per-hour wind, and existing fuel moisture—it should have been spreading at a rate of about 100 yards per hour.

Superheated pines exploded. Deer, fleeing the flames, ignored me as they bounded across the road. Rabbits, too.

Just after two o'clock, an old Navy TBM from Carson City made the first air drop of 600 gallons of bentonite slurry on the head of the fire. If this slowed its advance at all, it didn't for long. It did boost our morale.

So did arrival of the first substantial reinforcements. Before the fire was two hours old, fifty U.S. Marines from the Mountain Warfare Training Center at Pickle Meadows on the Sonora Pass road and sixty State of California inmates of the Inyo-Mono Conservation Crew led by CDF overhead were on the firelines. By three o'clock, over a hundred Marines were on the fire, and additional pick-up firefighters had been hired. And by four o'clock, when the air tanker's scout plane had reported the size of the fire to Fire Boss Hoag at about 250 acres, fire control personnel from the West Walker, Alpine, and Carson ranger districts had begun to arrive. A project fire overhead team left Reno about five o'clock, and overhead also began traveling to Bridgeport from the Central Nevada and Las Vegas divisions of the forest.

By late afternoon, the situation had begun to look like the project fire it had become. The fire was divided into four sectors, with Marion as sector boss over Forest Service crew bosses, Marine Corps line workers, and three bulldozers—"cats"—on the "hot" up-hill end of the fire. A PBY and two TBMs made a second series of air drops on the head of the fire. Much to my disappointment, I was assigned to set up the fire camp to receive men and equipment as they arrived. By nightfall, the camp and the night shift had been organized. I was eventually relieved as camp boss and told to rest up for a day shift job.

I've spent more comfortable nights than that one out in the sagebrush in a paper sleeping bag.

While I slept, the major battle of the campaign was fought at the top of Sawmill Ridge.

The plan was to take advantage of the cooling effects of the night air, which should slow the fire down, and stop it with a wide fireline at the top of the ridge. With seven cats and seventy-six men on the hot western sector, Marion was building that line across the ridge from north to south. At eleven o'clock, he tied in with the southern sector's line to contain the fire. But, two hours later, the wind shifted from the west to blow from the southeast and, as firefighters retreated for air and safety along their lines, the fire jumped the line. Another line was started to corral the blowup.

I awoke at first light to a transformed fire camp and landscape. The camp's population had doubled, at least, and there seemed to be plenty of fresh firefighters for the day shift. A mobile fire weather forecasting unit, complete with spinning anemometer and other instruments, had arrived and was in business. Best of all, a field kitchen was offering breakfast. Dirty, cold, and sore, I ate.

I was assigned to the northern sector as a crew boss, and given a Marine Corps platoon as a crew. The platoon's sergeant, probably figuring this young Forest Service fellow for something akin to a second lieutenant, suggested I work the crew-platoon through him. I readily agreed. Since the sergeant and his Marines were green at fire fighting, I held a brief fire school to explain what we were going to do, how to use the tools, and safety before starting up the line. They worked out fine. About the only problem I had, as we built fireline toward the ridgetop, was keeping them from bunching up—from working too close to each other for safety.

Fire Boss Hoag declared the fire controlled at noon, just twenty-four hours after it started. But there was still plenty of mop-up work left.

Late in the afternoon, my Marines and I were relieved by a Forest Service crew from Idaho, and at the end of the day shift I was released to resume fire prevention patrol duties the next morning.

There were plenty of questions to answer during that July first patrol of the Twin Lakes area. Thousands of campers and residents had watched the fire consume 350 acres of brush and timber between Robinson Creek and the top of Sawmill Ridge, and everyone wanted to talk about it. I, of course, capitalized on their interest by pushing my fire prevention message. The fire that had so impressed them, we knew, had been man-caused. A couple little boys, the investigation determined, had built a "campfire" in the brush.

Although the fire camp was removed that evening and firefighters from other districts and forests were returned to their stations, mop-up continued for several days.

The smoke of the Robinson Creek Fire had barely cleared when, on July 9, Forest Supervisor Sack and Frank Dunning, his fire staff officer, conducted the district's previously-scheduled fire preparedness and readiness inspection. Although we were still mopping up the big one, they told Ranger Hoag and Marion they were pleased with what they saw—including, much to my relief, my fire prevention program. But it seemed to me I was doing everything a fire prevention guard should do well—except preventing big fires.

On the evening of July 21, as I was completing a Twin Lakes patrol, I sighted and extinguished the last smoke. It was a smoldering stump about a third of the way up the ridge. The Robinson Creek Fire was officially out. But, in a way, it would never be out for me. As I crunched my way down the charred slope, it seemed this greatest defeat of my fire prevention career would burn forever in my memory.

The district logged another half-dozen man-caused fires that summer. One of these, early in August, burned a log cabin under construction at the Virginia Lakes Resort. A construction worker, who had just oiled the inside walls, tried to start a leaky gasoline generator inside the cabin. It blew up. He got out a window. The cabin

The site of the Robinson Creek Fire of June 29 and 30, 1964, before and after the fire and in 1965 after the burn was planted to crested wheatgrass.

Fire Control Officer Marion Hysell and a couple firefighters mopped up the Glacier Lake Fire of 1964.

burned to the ground. Marion sent me and two fire crewmen in the pumper—not to fight a structure fire, but to keep the fire from spreading into the forest. We got there in time to mop up.

Three of us responded on horseback to a report of a fire at Barney Lake in the Hoover Wild Area late one afternoon. Half way to Barney Lake, the ranger station recalled us. The fire was in the wild area, all right, but at Glacier Lake in a cirque below the jagged spires of the Sawtooth Ridge at about 10,000 feet. We'd have to start again in the morning. And we did, this time on foot with Marion leading the way up Blacksmith Canyon to the alpine lake where a fisherman's abandoned campfire had burned half the stunted whitebark pines ringing the shore. We put a line around it and mopped it up.

All firefighters like details to fires on other districts and forests. They get to see some different country, make a little money, and usually don't have to stick around for mop-up. The small Bridgeport District fire crew didn't travel much in those days, but four of us did go to the West Walker District one mid-August day for the forty-five acre Risue Canyon Fire in the Wellington Hills northeast of Coleville. Since, for some reason, no Forest Service rig was available for the trip, we four firefighters and our gear arrived at the fire shoehorned into my little red Volkswagen bug.

A couple days after the Risue Canyon Fire, the Bridgeport District's fire danger was rated extreme. This ended days off, put us on fire standby duty after work, and kept us from being detailed to large fires on the region's other forests. A few small fires later, the fire season ended.

One evening early that August of 1964, news of what came to be known as the Tonkin Gulf Incident penetrated the static of the little plastic radio Dick had left in the bunkhouse when he left for U.S. Army service. According to the report, North Vietnamese motor torpedo boats and gunboats had attacked a couple U.S. Navy destroyers operating in international waters off their coast. In response, President Lyndon B. Johnson had ordered carrier-based air strikes on North Vietnamese targets and Congress had passed a joint resolution approving use of force "to repel any armed attack against the forces of the United States and to prevent further aggression" in Southeast Asia.

"Here we go," I murmured, pondering our futures.

"Huh?"

"Never mind."

The next month, after I'd returned to San Jose State College—I'd left Berkeley in 1963 for study conditions more salubrious than I figured would be afforded by the impending violence that indeed erupted from what became known as the Free Speech Movement on that campus—I heard news I hadn't heard on the little plastic radio. Early in September, Congress had passed and President Johnson had signed the Wilderness Act of 1964 "to secure for the American people of present and future generations the benefits of an enduring resource of wilderness." To implement this policy, the Act set forth the concept of "wilderness" protected by law, established the National Wil-

As a result of the Wilderness Act of 1964, "I'd left the Hoover Wild Area..." at the end of the 1964 season. "Next spring I'd return to the Hoover Wilderness."

derness Preservation System, provided management guidance for that system, and provided for growth of that system.

With passage of that Act, fifty-four units on the National Forest System that had been administratively designated as wilderness areas by the Secretary of Agriculture or wild areas by the Chief of the Forest Service had become wildernesses set aside by Congress to remain "wild by law" forever. Among these original or "instant" wildernesses was the Hoover Wilderness. I'd left the Hoover Wild Area earlier that month. I'd return to the Hoover Wilderness the next spring.

I returned to the Bridgeport Ranger Station in the middle of June 1965 to news that Ranger Hoag was soon to transfer to the Sawtooth National Forest in Idaho. After a few days, and a going away party for Bob and Mary at Nevada Beach on Lake Tahoe—I rode up and back with Range Conservationist Ken Genz, his wife Sherry, and Pam, the new district clerk—they were gone.

"Through some arrangement, U.S. Marine Corps helicopters... flew us to several of those fires."

The fire season of 1965 began with a dozen small lightning fires. Through some arrangement, U.S. Marine Corps helicopters operating at the Mountain Warfare Training Center at Pickle Meadows on the Sonora Pass road—training, apparently, for Southeast Asia operations—flew us to several of these fires. Those brief flights saved us a lot of walking and helped us keep these fires small.

Sometimes the fires you don't go to can be harder on you than the ones you do. That's how the Wichman Fire that July turned out for me.

This one was a sleeper. It didn't show up until a day or two after the lightning storm. When it did, on the evening of July 9, Marion Hysell spotted the smoke rising behind the Bodie Hills while driving back to the ranger station from Twin Lakes. As soon as he got to the station—just as the telephones began ringing with reports of the smoke, he dispatched me with four crewmen in the pumper and followed in the Jeep with two others. Since we didn't know exactly where the fire was, the FCO sent me with the pumper crew toward the fire via the Masonic Road while he continued along the East Walker River highway to turn east onto Forest Route 28. Once the fire was pinpointed, he'd plan the attack.

"D-4-5, this is D-4-3, 10-20, over," the pumper radio crackled.

"D-4-3, D-4-5, just passing the Chemung Mine, over," I replied to the FCO.

"Les, I see the fire. It's burning in beetle-killed pinyon above the Wichman Ranch. Get up on New York Hill to radio relay for me while I scout it. I'll be out of radio range of the station pretty soon, over."

"Ten-four, D-4-5, 10-8," I complied. Soon we were up there, doing just that, listening to the rest of the talk, and watching the smoke rise from the flames as the sun set.

As darkness fell, Marion and Toiyabe National Forest fire staff officer Blaine Cornell who was scouting the fire from the air—operating under the philosophy that no fire is worth a man's life—agreed that a night attack on this fire, burning in remote rugged country of relatively low resource values, would be dangerous and impractical. They planned a daybreak attack that would involve extensive air operations. We were sent back to the ranger station to rest and prepare for whatever the supervisor's office cooked up for the next morning.

The initial attack on the fire was made early the next morning by 32 smokejumpers from Redding, California, and Medford, Oregon, soon aided by air tankers and helicopters and reinforced by Forest Service crews including one of 26 Sho Pai Indian firefighters from Owyhee, Nevada. By the time the 350-acre fire was controlled at seven o'clock that evening, these forces had been joined by—and, in the case of most of the smokejumpers, relieved by—hot shot crews from the Cleveland and Lolo national forests, a crew of Santo Domingo Indian firefighters from New Mexico, and additional Toiyabe National Forest firefighters. It sounded like one of the most exciting fire control operations the district had ever seen.

I say "sounded like" because I never got to the fire. Citing our district's high fire danger and low manning level, Marion had asked the supervisor's office not to strip the district of its own fire personnel. So, instead of going to the fire with Marion and two of our firefighters, I was back on patrol the next day.

As usual, Marion's call was a good one. Shortly after noon, as I was completing a Buckeye Canyon patrol, the pumper radio crackled with the Bridgeport Ranger Station's report to the Toiyabe National Forest's dispatcher of a fire at Stockade Flat. The fire, just west of Devil's Gate on the West Walker District, had been reported to the station by passing motorists. Range Conservationist Ken Genz, acting as district ranger between Ranger Hoag's transfer to the Sawtooth National Forest in Idaho and arrival of a new district ranger from Utah, had sent our two-man trail crew, recalled from the Hoover Wilderness when the Wichman Fire took off, to check it out. The fire was burning in Jeffrey pine timber and threatening to run. All they could do was throw dirt and call for help. I was already rumbling back to the ranger station when Ken called.

"You and Crawford will have to take a couple more guys up there and handle it. I've requested an air tanker."

Maurice Crawford assumed the duties of fire boss and, handing me the air net radio, told me to handle air operations.

I cranked up the pumper and all six of us built fireline for all we were worth. The air tanker, another old TBM, responded to my request for a good drop across the head of the fire. We held the Stockade Flat Fire to about five acres, and were released when Alpine and West Walker district firefighters—not so fresh from our Wichman Fire—arrived to mop it up.

Meanwhile, the Wichman Fire had been controlled, and the Bridgeport District fire crew was to relieve the remaining smokejumpers and hot shots being released the next morning. I didn't join this mop-up operation. Instead, sent back up New York Hill to provide radio relay services, I saw the sun rise and traverse a cloudless sky.

At first, there was almost something to that communications job. But toward noon, as radio relay between the fire and the ranger station began to seem a thing of the past, I got bored. Then more bored. And hot. All around, the horizon shimmered in the noonday sun. Eating lunch didn't help. Fighting boredom and fatigue, trying to stay alert, I began to botanize within earshot of the radio that, except for responses to my hourly radio checks, no longer emitted anything but static. Having exhausted my limited knowledge of New York Hill's equally limited flora, I climbed back into the cab to await messages to relay that never came. The activity of the last couple days cropped up on me.

I honestly don't know if I dozed off or not, but the lack of engine idle and radio static noise suddenly got my attention. Darn! The engine wouldn't crank over. And to make matters worse, I had failed to park facing downhill for a roll start. What a mess! Keep up like this, I told myself, and you'll be walking down the road...literally and figuratively. There was nothing between me and the East Walker highway but miles of dry, rocky, winding, seldom-traveled road out of the sagebrush and scrub of the Bodie Hills. Oh, well, at least it was mostly downhill. Grabbing the canteen and cursing my stupidity, I locked the truck and shambled toward the road and a long, hot walk out. It was three o'clock.

About the only thing I did right that afternoon was pick the right time to do things wrong. Not more than a minute after I reached the road, a Volvo station wagon rolled up behind me. Fool's luck. That was the first vehicle I'd seen on that road all day. Flagging it down, I sheepishly explained to the driver—a vacationing Los Angeles police detective—and his wife that my truck had "given out" and I could use a ride to town. Sure, they were going to Bridgeport anyway, and would be happy to drive me to the ranger station.

Forty-five minutes later, one embarrassed radio relay man was being bailed out of his predicament by the district's maintenance foreman, who drove him back up New York Hill with jumper cables and, just in case, a new battery. Soon he was back on the air. Later he was relieved to find out he hadn't missed any transmissions. Almost as important, he wasn't ribbed about the incident.

The middle of July brought heavy rains and—after 13 lightning-caused fires on the district—an early end to the fire season. Low to moderate fire danger ratings permitted a normal five-day work week, something I hadn't experienced before. I welcomed the chance to explore the eastern Sierra beyond the Bridgeport country, and was

"I was lucky to have Lloyd Parratt to travel with that summer. In addition to wilderness trips, Lloyd and I visited the White Mountain's bristlecone pine forest...."

lucky to have Lloyd Parratt to travel with that summer. In addition to wilderness trips, Lloyd and I visited the White Mountain's bristlecone pine forest where we met famous dendrochronologist Dr. C. W. Ferguson, Yosemite National Park where he introduced me to Chief Park Naturalist Douglas Hubbard, the Devil's Postpile National Monument, and the new Eldorado National Forest visitor center at Lake Tahoe. To say Lloyd taught me a lot would be an understatement.

While in Yosemite Valley with Lloyd, I forked over a couple bucks—a lot of money for a paperback in those days—for a copy of John W. Bingaman's 1961 book *Guardians of the Yosemite* about the park's first rangers. One afternoon a week or so later, while I was staffing the office, an old gent walked in and introduced himself as Bert Sault. "I know who you are, sir! One of Yosemite's first rangers! I've just been reading Bingaman's book!" Just over eighty years old, he'd begun as a Yosemite ranger in 1922—six years after the National Park Service was founded—and for years had patrolled most of the park on horseback. He seemed pleased I'd recognized him, and told me a couple stories about the early days.

One evening, on the way into Bridgeport for dinner, Lloyd confided in me an even more amazing story. He was working with a *Life* magazine writer on an article about his youngest son, Smitty, who on July 18, 1960, had been attacked and almost killed by a grizzly bear in Glacier National Park. Lloyd, a seasonal ranger-naturalist there since 1946, his wife Grace, and their three sons were spending another summer in the park. Ten-year-old Smitty was on a hike with two off-duty seasonal rangers. A couple vacationing Swedish schoolteachers joined them on the trail. Rounding a curve in the trail, the group encountered a huge female grizzly with two cubs. The four adults climbed trees; two were painfully clawed and bitten. Unable to get away from the bear, Smitty was horribly mauled. In critical condition for weeks, he survived to face years of reconstructive surgery. Congress had passed a bill in 1993 to permit the Parratts to bring suit against the government to help cover medical expenses, and at the time Lloyd told me this story, the pre-trial hearing was about a month off. *Life* published the story in its August 27, 1965, issue. Coverage of the Watts riots knocked Smitty's story off the cover.

The end of July brought a new district ranger, E. Lynn Mitchell, his wife and sons. A graduate of Utah State University, he had been district ranger on the Tooele Ranger District of the Wasatch National Forest.

August brought more rain, and the threat of floods replaced the threat of fires. A flash flood washed through the district's Alum Creek Campground about a dozen miles south of Hawthorne, Nevada. Buckeye Creek and Green Creek had the hydrologists worried so much that, in the case of Green Creek and just to be on the safe side, I was sent to warn campers and permittees of the possibility of flooding. It didn't happen.

My fourth Forest Service summer looked like my last. The war was heating up in Southeast Asia, and the Selective Service System

A new Bridgeport District forest ranger, E. Lynn Mitchell, arrived in July 1965.

was hot on my trail. I needed one more semester to complete college, but my four-year student draft deferment had expired. A hopelessly healthy young American male in his early twenties, I soon would be drafted.

Back in San Jose, as the fall semester began, I redoubled my efforts to join the Navy. Previous attempts had failed because I wore glasses, but at last I found a program that would consider me. Application processing bought me enough time to graduate in January and kept the draft board away. I was finally sworn into the Navy early in April 1966, to compete for selection to become something called an air intelligence officer. It was a long shot, I was told, but I was selected. Since there wouldn't be a place for me in the Aviation Officer Candidate School, Pensacola, Florida, until late 1966 or early 1967, I was free to return to the Forest Service.

My letter to Ranger Mitchell brought an immediate reply. They needed me, he wrote, "yesterday."

I returned to the Bridgeport District for the last time in April 1966, for what proved the most eventful of my Toiyabe patrol summers.

Chapter Five: The Summer of '66

The summer of 1966, my last on the Toiyabe National Forest, proved a fire season to remember. I had arrived in April after graduating from San Jose State College in January and working as a substitute teacher for a couple months while my Navy future was being decided, and the fire season began in May. By the beginning of June, long before the summer crew had arrived, the Bridgeport Ranger District had counted ten wildfires.

Most of those early fires occurred during one late May week of lightning storms over the drier pinyon and juniper hill country. They were generally harder to get to than they were to put out. As the two fire control people then on the district, Marion Hysell and I handled most of them. After the first couple, reported by telephone, had snatched us from our dinners and kept us out late, we knew what was up when we heard the ranger station's three rings in the evening. Going to those fires became as routine—but not nearly as restful—as going to bed. We were, in Marion's words, "darn near give out" by the time that string of fires ended.

By the end of that week, something like twenty-two acres had burned. One fire, a BLM fire near the Chemung Mine in the Bodie Hills, accounted for all but one of those acres. We controlled that fire with the help of air tankers and a Job Corps crew from Carson City. Two others were quickly suppressed for us by a West Walker Ranger District helitak crew, and two others were put out by fire suppression cooperators. Of course, I spent several hours in the office writing the fire reports.

The Bridgeport Ranger Station in 1966.

Earlier in May, before that spate of fires when Marion and I were getting the district's fire equipment ready for the fire season, we discovered the slip-on pump unit on the fire rig needed some repairs. So, late the next Monday morning, I drove the 113 miles north on U.S. Highway 395 through lightning, thunder, rain, and hail toward Reno and a thoroughly aggravating day.

I arrived in Reno, delivered the rig to the mechanic at the Toiyabe National Forest warehouse, and went down the street for a hamburger. By the time I returned to the warehouse, the mechanic had decided to install another pump unit on the rig and keep the original pump unit in the shop for overhaul. He wanted to do some work on the truck, too. I had several hours to wait.

After an hour or so during which I gathered some equipment the district needed, the supply clerk gave me a station wagon and some errands to run around town. I was happy to be useful. Finally, about half and hour before quitting time, the mechanic told me the rig was ready. I loaded the equipment, radioed the ranger station that I was leaving Reno, and headed south toward Bridgeport. I didn't get far.

In southern Reno, at the intersection of Kietske Road and U.S. 395, the truck lost all power. I managed to coast it onto the shoulder of the road and stop safely, but the radio was dead and I couldn't call the dispatcher for help. The truck battery that was also attached to the pump unit seemed to have shorted out. I couldn't tell for sure. Stranded, I hitched a ride to a gas station, telephoned the dispatcher, and caught him on the way out the door. He said he'd have the garage crew on the way to help me, "pronto."

I hitched a ride back to the truck. Before long, a southbound Forest Service pickup pulled off the road. It was Malcolm Cockerham, a most affable Carson Ranger District forester from Mississippi on his way home from a tree planting project in Dog Valley. He helped me push my dead truck across U.S. Highway 395—four lanes of traffic eager to get to the slot machines and crap tables in Reno or back to the smog in Los Angeles—to a safer place to wait, then stuck around until the garage crew arrived with the tow truck.

Back at the warehouse, the mechanics decided my truck needed "a little more work" and would not get me back to Bridgeport any time soon. So, they quickly mounted the pump unit on another pickup—without a radio or flashing red lights—just "completely overhauled" by Job Corps trainees, and promised to get my rig back to Bridgeport in "a couple-three days." It would take that long because, in the process of removing the pump unit from my truck, one of the mechanics put his rear end through its rear window and added another repair job to the list. In a great leap of faith, I loaded the other equipment on the just-overhauled loaner and by six o'clock was chugging, clanking, rattling, and grinding toward Bridgeport.

My faith was neither justified nor rewarded. The loaner was low on fuel, so I stopped in Carson City for gas. When I tried to leave, it wouldn't start. Charging the battery didn't help. Now, I'm no more a mechanic than I am a buckaroo, so had to accept the gas station attendant's diagnosis of "a bad solenoid" along with his help starting the engine with a screwdriver. Needless to say, once that worked, I didn't turn that rig off until I got back to Bridgeport. Luckily, the guys at the State of California agricultural inspection station at Topaz Lake just waved Forest Service vehicles through without stop-

"Most of my 'fire prevention patrols' that spring were dedicated to sprucing up the district [and] posting fresh Smokey Bear signs."

ping. I finally arrived at the ranger station sometime after ten, told Ranger Mitchell my tale of woe, and fell into bed.

Eventually, the mechanics delivered my patrol rig to the ranger station, remounted the pump unit on it, checked it all out, and took their "completely overhauled" pickup back to Reno. I was ready to patrol.

Most of my "fire prevention patrols" that spring were dedicated to sprucing up the district for the summer. In addition to posting fresh Smokey Bear signs, this included repairing and repainting signs and other structures.

One afternoon, while working alone at the Huntoon Campground just a mile north of the ranger station, I carelessly gashed my knuckles while replacing the lid on a five-gallon stain can. As red blood began to flow, I wrapped my right hand in a handkerchief—Mom told me a gentleman always carries a handkerchief, and I was being a gentleman that day—and sped south to the ranger station for some first aid.

As I turned off U.S. Highway 395 into the station, I saw two women walking from their parked car toward the office. It was Saturday, and the office was closed. But it seemed wrong to ignore them.

"We need a campfire permit," they responded when I asked how I might help them.

I unlocked the office, and all three of us managed to ignore the blood seeping through the handkerchief as I issued the permit.

I locked the office as they left. Beginning to feel a bit dizzy, I walked over to the FCO's house to get some help for my knuckles. I knew it looked worse than it really was, but that didn't seem to help much. I was let into the kitchen through the back door and, while Marion was getting his first aid kit, began washing my wound. The next thing I knew I was sitting on the kitchen floor with several wide-eyed Hysell kids staring at me. It was embarrassing. I had never fainted before, and have never fainted since.

Half an hour later, my right hand dressed and bandaged, I was back at work.

My early arrival that year also got me the job of presenting the district's spring fire prevention programs at the local schools. While my program was well received by the Bridgeport Elementary School's fifty or so students, I never felt more appreciated than during the visit to the one-teacher school in the rundown railroad town of Mina, Nevada. It was a hot, dusty, all-day trip to talk with and show a couple movies to a small group of children, but it was worth it. The warm reception I received, the rapt attention I was paid, and the gratitude expressed by the teacher and the students were enough to make me wonder if anyone else had ever visited that school.

The early fire season continued. One June afternoon, a lightning storm ignited a blaze in timber and brush on the moraine south of the Twin Lakes. With the help of an air tanker, which put on quite an air show as it pulled up over Lower Twin Lake after applying retardant on this difficult target, we controlled the fire by midnight.

Early the next morning, I had to drive over Sonora Pass and across the Central Valley to the naval air station at Alameda on San Francisco Bay for a scheduled interim Navy physical examination. For some reason, unimpressed by my protestation that I had to drive back over the Sierra that evening, the medical officer dilated my eyes.

After dinner at a college friend's home, I set off for Bridgeport—fearing the California Highway Patrol would notice me driving at night with dilated pupils covered by sunglasses and think I was on something. The oncoming headlights were blinding. I made it as far as Sugar Pine, where for four dollars I holed up in an empty ski dorm. Just west of Sonora Pass the next morning, on what had to be the steepest and narrowest stretch of California Highway 108, I had a flat tire. I got to Bridgeport too late for breakfast, but just in time for more fires.

Five fires were ignited by the lightning storm that built up quickly during that afternoon's lunch hour. I had gone into Bridgeport with Mark, the district's Hoover Wilderness patrolman—and Lloyd's eldest son who had several summers as a fire guard in Glacier National Park under his belt, for a hamburger at the Purple Onion drive-in. We had just finished when the lightning started, and were hurrying back to the ranger station when we spotted the first plume of smoke near the mouth of Buckeye Canyon. The FCO dispatched us to the fire with the pumper. Burning in second-growth Jeffrey pine, this was one of those rare fires we could drive to. Mark and I had been at the fire barely long enough to contain it with a scratch line and wet down some hot spots when the radio crackled with word of another smoke.

"D-4-5, this is D-4-3, over." It was Marion Hysell in his Jeep.

"This is D-4-5, over," I responded.

"Les, I'm on my way to a fire on the north shore of Upper Twin Lake, just east of Mono Village. I've got air tankers and a Job Corps crew from Carson City on order. Wrap up your fire as quick as you can and join me, over."

"Ten-four. We'll be there as soon as we can. D-4-5, 10-8." That fire, I knew, was close to a few houses.

Just as I hung up the mike, a pickup pulled up. Two hands from the Hunewill Ranch, which had land and cattle in the canyon, had been attracted by the smoke. What a break! I quickly hired them on fire time slips, gave them a pulaski and a shovel, and left them to finish the job.

The Mono Village Fire of 1966 "spread outward from its point of origin in an almost perfect 'ring of fire' through scattered Jeffrey pine and brush."

After it was ignited by lightning, the blaze that would be named for the nearby Mono Village resort had spread outward from its point of origin in an almost perfect "ring of fire" through scattered Jeffrey pine and brush. By the time Mark and I had arrived, the FCO had organized a couple small crews of pick-up firefighters. Supervised by Bridgeport District personnel, they were working their ways up the fire's flanks.

I parked the pumper—with red light flashing—on the highway next to the lake. A hose-lay from a portable pump on the lake's edge and through the pumper would move plenty of water up the slope to the fire. About the time we began putting this water on the fire, the air tanker checked its upslope progress with a retardant drop. We held the fire to about ten acres. The Job Corps crew arrived in time to mop up.

Only later did it occur to me that we had been able to drive to and put water on—rather that walk to and throw dirt at—two fires in one day!

Later that evening, district firefighters were dispatched to three more smokes.

Close on the heels of these fires, the Toiyabe National Forest experienced one of its larger forest fires of the era. Up on the Carson Ranger District, the Galena Creek Fire was ignited—by a discarded cigarette, according to the investigation report—about a dozen miles south of Reno around noon on June 27.

The fire was discovered and reported by a Nevada Division of Forestry patrolman. Under a cooperative agreement, the state agency patrolled an area that included this national forest land. The patrolman, in a pumper unit with one other man, attacked the fire about twenty minutes later. But they couldn't contain the blaze, which was spreading quickly toward the north and east in timber stands that included the rare Washoe pine, and the Forest Service started to roll reinforcements.

Less than an hour after the initial attack and just as the Forest Service was assuming responsibility for combating the erratically spreading flames, the fire blew up and crossed Nevada Highway 27. All traffic on this heavily used road was stopped.

A new Forest Service tractor-lowboy rig, transporting a bulldozer for fireline construction, was trapped on the highway by the fire it was to help control. The truck driver and cat skinner escaped, but the rig and cat were destroyed. The Toiyabe had a real project fire on it hands.

Fire Boss Blaine Cornell, from the forest supervisor's staff in Reno, sized up the situation. A spot weather forecast called for strong down-canyon winds and low humidity through the afternoon. He planned to stop the fire's rapid eastward spread toward homes and ranches first, then control its mountainous northern and western flanks. Manpower and equipment were ordered for a four-sector fire, and that's where I came in.

When the supervisor's office requested two qualified crew bosses from the Bridgeport District for the night shift, Ken Genz and I were dispatched to the fire. It was dark by the time we first saw the fire—hellish orange flames burning along a two-mile front atop a ridge off to the west—from the Washoe Valley. This would be the biggest fire I'd ever fought. We turned off U.S. 395 and made our way up Nevada 27 to find the fire camp.

Ken and I got to the fire camp just in time for night shift assignments and too late to eat. We gulped black coffee during the quick briefing around a map board. By then, we were told, the fire's eastward run had been stopped by a backfire from Callahan Road and further contained by firelines on the northern and southern flanks as far as the highway. The night shift would build and hold a three-mile line around the remaining northern and western flanks. Two line locators were needed, and we were them. Our job was to lead cats up both flanks to head the fire off during the cool night hours. Crews with hand tools would follow the cats at a safe distance, improving the lines that would link up along the top of the ridge and corral the fire by morning. Except for the cats Ken and I would be leading, we'd be on our own.

After the usual delays inherent in working with heavy equipment, I was leading my cat up the fire's northern flank.

The night was dark, but the fire on my left provided so much light I didn't need the headlamp on my hard hat. Trees continued to torch off but, as expected, the fire was now advancing more slowly. I soon had the hang of nighttime line locating, and worked the cat as close to the fire as I could while avoiding heavy timber that would slow it down. I put all I knew—and all the cat skinner knew—about fire fighting into the job, and we made steady progress along the fire's flank and around its head as the night wore on.

As far as I know, I made only one mistake that night. But it was a big one, and I came close to paying dearly for violating a cardinal rule. Carelessly allowing myself to get downslope of the cat, I suddenly gained a whole new perspective on boulders.

Just after dawn, Ken and I tied our cat lines together, pinching off the head of the fire. If not yet controlled, the Galena Creek Fire at least was contained. The sector boss soon arrived. Fresh crews were on their way up the mountain to relieve the night shift, he said, and we were to return to the fire camp. Food and rest at last! Well, almost. First we had to get there.

Wanting to see the rest of the fire, I opted to walk off the mountain along Ken's cat line. The effects of the fire's rapid and erratic spread were obvious. While most of the area inside the fireline was blackened, some stands of timber had been bypassed and were as green as ever. By the time I hit the state highway, which I would follow for about three miles to the fire camp, the sun had chased the chill from the morning air. I met fresh crews on their way to the western flank. After a while, I came across the remains of the trapped tractor-lowboy and cat rig. A nice new outfit only a day before, it was all blistered metal and burned rubber. I walked on through yesterday's cool, green forest—now a charred and smoldering wasteland.

About that time, at ten o'clock in the morning, the Galena Creek Fire was declared controlled at just over 1,200 acres. Two homes had burned.

Thirty minutes later I drifted into the fire camp. Hot and dirty, hungry and thirsty, and worn out, I craved a cool shower, something to eat and drink, and sleep. Knowing the shower was out of the question, I made my way through organized chaos to the mess tent. Too late for breakfast, I washed sandwiches down with coffee. Ken and I were to stick around for the night shift, we were told, so we had to bed down at the fire camp. I have always had a hard time sleeping in the sunshine, even when tired. About the time I finally did drift off, I was awakened by the hustle and bustle of a Reno television news crew filming an interview with Fire Boss Cornell right where I was trying to sleep. It was a thoroughly miserable day.

Once under control, project fires have a way of getting real boring real fast. This one was no exception. By the time I reported for the night shift, the Galena Creek Fire was being mopped up. Last night's initial attack line locator was tonight's mop-up squad boss. It was a long night. But, because the Bridgeport District was having so many fires that spring, Ken and I were released the next morning.

We left the fire camp after breakfast, and headed south on U.S. 395. It was hot and we were tired. Before long, just to keep ourselves awake, we needed a break. So, feeling more than a little conspicuous, two smoky, sweaty, and unshaven guys in green shambled into a Carson City ice cream parlor to down banana splits and several cups of coffee. It was a refreshing break, but not a patch on the shave, shower, and sleep to be had in Bridgeport.

As the summer wore on, the country dried out. By the middle of July the fire danger was extreme, and Ranger Mitchell announced the Toiyabe National Forest's prohibition of open campfires outside developed campgrounds and other fire safety restrictions. This required special signs and increased the intensity of my patrols.

The requirement for all forest visitors and users traveling off paved roads to carry fire tools—a "shovel, ax, and bucket per car or pack train"—brought the Bridgeport General Store a brisk trade in those items.

MONO COUNTY'S PIONEER NEWSPAPER — ESTABLISHED 1863

WELCOME! EAST HIGH SIERRA

BRIDGEPORT

Chronicle-Union

OFFICIAL NEWSPAPER OF THE COUNTY OF MONO

Our 103rd Year 10c PER COPY P. O. Box 218 BRIDGEPORT, MONO COUNTY, CALIFORNIA ALL RIGHTS RESERVED Friday, July 15, 1966 NO. 28

Reichert in Mono Race As Write-In Candidate

Doris Reichert of Benton this week announced that she will run as a write-in candidate in the special election Aug. 9 for the post of Mono county supervisor, District 2.

She ran for the office in 1964, and was defeated by one vote by Frederick B. Dawson, who now holds the position which is the subject of the Aug. 9 recall election.

She is now in her 20th year of residency in Mono county; has been on the Mono county board of education for 14 years, and is a past chairman of the board. She has been active in numerous civic activities in Mono and the Benton area.

Also on the ballot in the Aug. 9 recall election are Howard Emsminger, Jack Partridge, and John Fobes, all of Long Valley.

NO CLOSURE SEEN

In Bishop, a spokesman for the Inyo National Forest said that no campfire closure is foreseen in any of the forest's districts, even though the fire danger is also severe there.

Toiyabe Forest Closed To Open Campfires; Fire Danger Extreme

BRIDGEPORT — All areas of the Bridgeport District, Toiyabe National Forest, are scheduled to close to open fires and smoking effective at noon today.

The closure, effective in all districts of the Toiyabe National Forest, was announced locally by E. Lynn Mitchell, district ranger.

It was authorized by Ed Maw, forest supervisor, in Reno.

Mitchell said the closure, which affects all open fires except those under special permit, is being made because of extreme fire danger in the forest.

The current fire season has seen the Bridgeport district hit by 23 fires, only five of which were man-caused. That figure, however, means the local district "has been hit as heavy as any," Mitchell said.

Mitchell said two additional steps could be taken if the fire

Swift Action Douses Mammoth Lakes Fire

MAMMOTH LAKES — In a swift action coordinated by officials of the Inyo National Forest in Bishop, 125 firefighters prevented a Wednesday afternoon blaze from endangering both local residents and U.S. Highway 395.

The fire, discovered in a brush-covered area near the Mammoth Lakes dump, was blamed on sparks from the dump.

Winds estimated at 25 mph, with gusts up to 40 mph, spread the flames rapidly, but fire suppression crews were on the scene in minutes.

John Harmening, fire control officer of the Mammoth District, Inyo National Forest, who discovered the blaze, said the first crew was battling the flames within 5 to 10 minutes.

Within two hours, over 100 men were present, and the fire was declared under control at 5:30 p.m. when a ring of firefighters, using mostly hand tools, managed to circle the area.

Aircraft Help

In addition, U.S. Forest Service aircraft from Fresno, Porterville and Chino dumped solutions of

Lions Busy With

"As the summer wore on, the country dried out. By the middle of July the fire danger was extreme, and Ranger Mitchell announced the Toiyabe National Forest's closure to open campfires."

I was on patrol in the Twin Lakes area early one August afternoon. It had been a slow morning and, just as I began to wish the pace would pick up, it did. But not in any way I expected.

"D-4-5, this is KMB-652." It was Pam, the district clerk.

"KMB-652, this is D-4-5, over."

"Les, we've had a couple reports of a dead sheep in the Honeymoon Flat Campground. Will you check it out, please."

A dead sheep? I hadn't seen sheep in the Twin Lakes basin for days. This one, I surmised, must be a stray from a Summers Meadow allotment. Oh, well, all in a day's work. Hoping the poor critter wouldn't be too ripe, I responded affirmatively. Guffaws from 652.

A few minutes later, I approached Honeymoon Flat. The sheep wasn't difficult to locate. Lying just off the highway, on the road into the campground, it appeared to have drawn more people than flies since its demise. The small crowd's attention shifted to me as I pulled up alongside the woolly carcass of a fully-grown ewe.

The extreme fire danger of the 1966 season required a full range of wildfire prevention precautions.

"Howdy," I said to no one in particular, closing the patrol truck door behind me. A cool afternoon breeze blew off the Sierra and stirred the aspen.

"It's a dead sheep!" explained a helpful bystander. I nodded agreement and smiled thanks for his assessment. "Are you going to get rid of it?"

"Yes, sir." Scratching my head briefly, I sized up the problem and then, in reluctant resignation, strode purposefully toward part of the solution. The crowd's eyes followed me intently.

Fortunately, the campground garbage had been collected recently, and I found an almost empty can. Quickly transferring its contents to another, I hauled the can back and pushed it under the ewe. Because dead animals don't exactly leap into garbage cans, I had to pry this one into its casket with a shovel. Once the animal was in the can, I set it upright and slapped on the lid. Then, dropping the tailgate to the horizontal position, I dragged the canned sheep to the rear of the patrol truck. There was just enough room for the can to ride behind the pumper unit and the fire tool box.

But one problem remained. That was one heavy garbage can! I doubt I could have lifted it, at least not gracefully. But an independent streak doesn't allow me to ask for help—even in situations like that. With all eyes on me, I stooped to pick it up....

"Here, let me give you a hand," volunteered a burly fellow. Saved!

"Thanks."

We easily lifted the can onto the tailgate, where I lashed it securely in place. Having a dead sheep fall off one's patrol truck onto the highway, it seemed to me, would be most embarrassing. With a ripple of applause and a few chuckles, the crowd dispersed. And the sheep was soon disposed of in the nearby Forest Service dump I would burn in a few days.

My patrol route that day continued around Sawmill Ridge, crossing and recrossing Buckeye Creek as it doubled back toward Big Meadow through open stands of Jeffrey pine carpeted with sagebrush. The reddish bark of the pines blazed against a backdrop of fused greens and blues in the late afternoon sunshine. Dark shadows had already captured

Flatiron Ridge, and soon would advance across Buckeye Canyon. The wind was whipping up, carrying the pungency of the dry forest.

"All stations, this is KMB-400 with the weather." The voice of Waldo VanArsdale, the Toiyabe's veteran dispatcher, suddenly boomed over the radio to remind me of how late in the afternoon it was. The fire weather forecast, always at four o'clock, was the same as it had been for weeks: hot, dry, and windy. Additional fire safety precautions—even closing the forest—were being considered.

I hoped to finish the patrol and get back to the ranger station for a beer and taco party that evening.

At the Buckeye Canyon roadhead, I noticed a campfire flickering next to an apparently unoccupied lone vehicle—a pickup with a large camper body. Occasional sparks blew from the fire into nearby duff and brush.

Clambering out of the patrol truck cab, I took a quick look around and concluded that, unless someone were inside the camper, extinguishing the campfire would be my job. I knocked on the camper door and waited. There was no response. Only as I turned to begin putting out the fire did I hear a reply.

"Just a minute, please," a woman's voice trembled. There was a rustling inside the camper. I turned back.

The camper door swung open and I found myself staring into the barrel of a small .22-caliber pistol grasped in a trembling hand. The woman, clad in a bathrobe, looked me over furtively and then, obviously relieved, gasped, "Oh, I'm sorry! I didn't notice your uniform! My husband has gone fishing, and I'm here all alone."

I continued to stare, transfixed by the weapon in her hand. Then after a few moments—which dragged on like eons—and another "I'm sorry," she smiled, glanced down, *uncocked* and lowered the pistol.

I blinked, swallowed hard, and managed a few words.

While I don't remember exactly what I said to the woman, I do recall that my words were very reassuringly polite and that they reflected my profound concern for her safety. The pistol was never mentioned—as if I were accustomed to being greeted at gun point and hadn't even noticed. But I wasn't. And I had.

Bridgeport Ranger District Fire Prevention Guard Les Joslin at the end of a long day on patrol in 1966.

The dryness in my mouth and the pins-and-needles sensation I felt as I drove off betrayed my realization of just how close I'd come.

"D-4-5, this is 652, over." It was the FCO. I responded, realizing I had missed my five o'clock radio check. "You'd better get back in here...that is, if you think you can handle beer and tacos after your sheep patrol!" He chuckled. So did I. He didn't know the half of it!

Despite the extreme fire danger, we had only two more small man-caused fires that summer. Either fire prevention was working, or we were darn lucky. Or both. Lightning-caused fires, of course, could not be prevented.

The thunderheads that built up over the Sweetwater Mountains north of the ranger station late one August afternoon looked particularly black and ominous. Appraising the clouds through the office window, Marion Hysell knew that country was in for a full dose of nature's fury, and soon.

"Com'on, Les. Let's take a ride up the road." He had sized up the coming storm and figured the best place to watch for the lightning strikes it would bring. I left the fire report I was writing on the desk and followed him out the door.

We parked the FCO's Jeep on the shoulder of U.S. 395 at Rattlesnake Point. The wind had quickened, and seemed to object to my efforts to spread a map on the Jeep's hood as it blew into the gathering storm. Suddenly, lightning lit up the sky, and thunder boomed across the Sweetwaters and echoed through the canyons. For the better part of an hour, lightning stabbed at the landscape. Marion winced as he watched a particularly persistent bolt strike northeast of us. A deafening clap of thunder about ten seconds later placed it about two miles away, probably on the western slope of Mack Canyon.

"Holy jumpin' coyotes!" the FCO muttered knowingly.

As the storm subsided, we wondered how many sleepers would show up. There were no smokes yet. But only a few large drops of rain had fallen where we stood, and probably not much more in the mountains.

"We'll fly the district in the morning," Marion decided. Then, pointing in the general direction of Mack Canyon, he added, "We'll have a fire up there for sure. Want to be the one to find it?"

He was offering me the flight.

"Sure."

I was leaning against a pickup, district map book in hand, when the radio-equipped two-seat Cessna airplane, contracted by the Forest Service for aerial patrols following lightning storms, arrived at Bryant Airport from Carson City the next morning. The veteran pilot, who knew the country better than I ever would, asked where I wanted to go. Determined to find Marion's fire, I asked to fly the northern end of the district, including the Sweetwaters, first, followed by the Bodie Hills and Mono Basin country and then the Sierra.

We were soon aloft, climbing over Bridgeport Valley as we turned northwest toward Rickey Peak. As we flew over the ranger station, I made the usual radio check.

Within minutes I was searching the Sweetwaters for smokes, but not finding any. After the plane had passed over Mack Canyon, I keyed the mike.

"Six-five-two, patrol plane, over."

"Go ahead, patrol plane, this is 652," Marion responded.

"I've just eyeballed the Swauger Creek drainage and Mack Canyon, and no smoke. Over Murphy Flat now, over."

"Ten-four, patrol plane. Take another good, slow, close look at that country before you head south, over."

"Ten-four, 652, patrol plane, 10-8."

Skirting the western slope of Mount Jackson, we turned to follow U.S. 395 northward and searched the several canyons draining the Sweetwaters to the southwest before flying up Swauger Canyon again, and again turning toward the southeast. Then, as the plane milled over the ridge between the East Fork of Swauger Creek and Mack Canyon, the pilot and I saw the smoke. "There it is!" we shouted in unison.

I tried to sound nonchalant as I called the station. "Six-five-two, patrol plane, I have a smoke, over."

"Ten-four, patrol plane. This is 652. Mack Canyon?" Marion's question sounded more like an answer.

"Right where you said it'd be."

The pilot was flying a tight circle around the smoke to give me a good look. I was happy I had taken two air-sickness pills before leaving the ranger station. "It's burning in a group of big old junipers...looks like only one tree...or snag...actually burning...on the western slope of Mack Canyon...about smack in the middle of section eleven," I reported, comparing the country with the map. "Fuel's pretty scarce, so I don't think it'll go anyplace fast. That old road up the mountain across Swauger Creek, the one that leaves the new road about a mile north of the highway, should get you there."

"Ten-four, Les. I'II start up that way in the Jeep with a couple fellows. Go ahead with the flight, but be ready to come back to give me a steer to the fire if I need it. Over."

He wouldn't, I thought, but I would.

About an hour later, having flown the rest of the district without spotting any more smokes, the pilot and I decided to see what kind of progress the FCO and his firefighters were making.

I soon saw the Jeep picking its way along the old road and raised him on the radio.

"Patrol plane, this is D-4-3. See any other smokes, over?"

"No, D-4-3," I responded. So far, it looked like the storm was letting us off easy.

"How close are we to the fire, over?" Marion couldn't see the smoke on the other side of the ridge. But I could see the Jeep and the smoke.

"Real close. Follow that old road to its end, and you'll be able to walk downhill a few hundred yards to the fire." The idea of walking *downhill* to a fire amused me. Could this be a first?

"Ten-four, Les. You may as well return to base."

Half an hour or so later, feeling the way flying usually left me feeling, I ambled into the ranger station office. Pam had just been on the radio with Marion, and he wanted me to load a bunch of five-gallon water packs into a pickup and drive up to the fire. The twelve-mile trip took over an hour, with most of that time spent negotiating the last couple miles up the old road—generously labeled an "unimproved dirt road" on the district map—Marion and his two firefighters had taken earlier. I parked the pickup next to the Jeep and, shouldering a couple of the water packs, picked my way down the canyon to help them mop up.

I'd learned early in my fire control career that, after detection and initial attack, everything else about fighting a fire was anticlimactic. Helping mop up those big old juniper snags didn't change my mind.

My last few Toiyabe patrol weeks coincided with California's mule deer hunting season and provided my first experience with deer hunters. In previous years, I had returned to college before the hunters had arrived. So the opening weekend of that hunting season was an eye-opener for me. I had no idea so many people came so far to hunt. But, as the motels in Bridgeport and the campgrounds and other camping spots all over the district filled with hunters and their rigs, I began to appreciate the fire prevention job ahead. As the first day of hunting season approached, the Bridgeport Ranger Station was mobbed by hunters wanting campfire permits—that was a good sign, I figured—and information.

As usual, Marion Hysell had planned for that onslaught. On opening weekend, he put his plan to work. As he flew the district in a helicopter, he vectored me, Ranger Mitchell, and a couple other district personnel assigned to patrol duties toward fire prevention "hot spots" on the ground. On the ground, in addition to preventing fires, we did the usual duties of Forest Service patrolmen during hunting season. Also, along with California Department of Fish and Game officers and Mono County sheriff's deputies, we validated the tags successful hunters attached to their game.

Forest Supervisor Ed Maw detailed members of his Reno staff to help district personnel during this opening weekend patrol effort. Mr. Al Hayes, the Toiyabe National Forest administrative officer, was assigned to ride with me.

"Les, I'm just an S.0. paper pusher," he joked as we headed up the Buckeye Road on opening day. "You're the expert here. You know the country and the job. Just let me know how I can help you."

I did. And he helped. By the end of opening weekend, Mr. Hayes and I had contacted what seemed like hundreds of hunters, validated countless tags, and put out more than a few abandoned campfires. I'm pretty sure we prevented some fires.

But the deer hunting season didn't end with the end of the opening weekend rush. It went on through most of October, and so did my fire prevention patrols.

Contacts with hunters were usually, but not always, pleasant. One day, not long after opening weekend while Marion and I were patrolling in Buckeye Canyon together, we came across a particularly drunk and obnoxious bunch. They had a particularly large, roaring campfire, and we stopped to do our duty. We did, but not without a barrage of foul-mouthed comments about "cops," which Marion explained

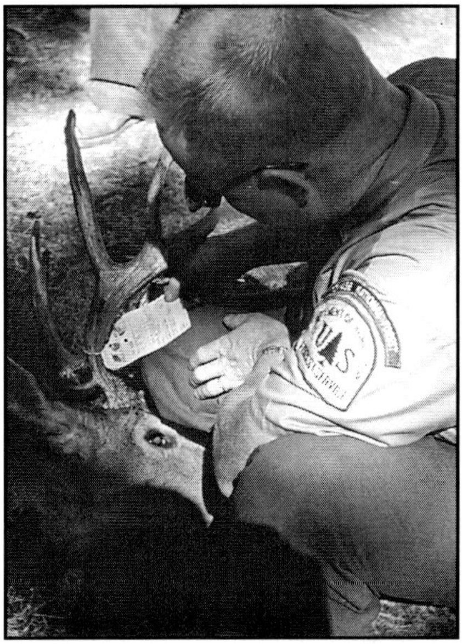

Al Hayes, the Toiyabe National Forest administrative officer, validated many deer during the opening weekend of the 1966 deer hunting season.

we were not. Half way through this encounter, I began to think seriously about the fact they were armed and we were not.

Just a few minutes later, in a nearby camp, Marion and I roused from a drunken stupor a hunter whose campfire was about to ignite the tent in which he was snoring the day away.

Trouble that hunting season came on four feet as well as two. One day, as I patrolled the Buckeye Road, I spotted a campfire burning in an apparently unoccupied camp on Buckeye Creek. I stopped the patrol truck, and got out to investigate. The campfire was attended only by a large German shepherd chained to a pickup bumper. I couldn't raise anyone, so I started to put out the fire.

That got the dog's attention. He growled in displeasure.

But he was chained, so I ignored him. I shouldn't have. Suddenly, the barking beast ran toward me. "How long is that chain, anyway?" I asked myself. But it was too late, and the chain was too long. Fangs bared, the charging dog leaped at me. As it did, I twisted to one side and kicked at it as I tried an all-too-slow exit. We both made contact. I caught the dog in the rib cage, and it sank a couple teeth into my upper arm before I got out of range.

"I especially enjoyed one sunny, crisp autumn morning when the two of us cleaned up litter left at Big Virginia Lake by the summer's fishermen."

As the dog strained at the chain, barking and growling as only German shepherds can, I retreated to the truck. I took off my shirt, and was surprised to see how little I was hurt. The wound was no more than a couple small punctures and some superficial scratches. Not much blood flowed. I quickly applied some disinfectant and a couple bandages, slipped back into my shirt, and set about writing the dog's owner a citation for abandoning his campfire.

Just as I got to the point I needed the fellow's name and address, he showed up. "What's goin' on. I heard barkin'."

"You left your campfire unattended," I informed him. Then, somewhat sheepishly, I added, "And your dog bit me."

"I just went up the creek a ways," he offered in weak defense.

"And the dog was watching your fire?" I commented before asking his name and address. I handed him the citation, and advised him I would contact the authorities in Glendale, where he lived, about quarantining his dog to make sure it didn't have rabies. When I got back to the ranger station, I did just that. I also paid a visit to Doctor Nichols in town.

As the hunting season waned, I spent several days working with Marion on a variety of odd jobs at the ranger station and around the district. I especially enjoyed one sunny, crisp October morning at Big Virginia Lake when the two of us filled many burlap bags with litter left by summer fishermen.

The other days I patrolled, contacting the remaining hunters and validating the occasional late-season deer. And getting used to the idea I'd soon leave.

My fire prevention career on the Toiyabe National Forest ended on a high note. Despite the high fire danger and the many deer hunters in the country—a potentially volatile hazard and risk combination, there were no human-caused wildfires on the Bridgeport District that hunting season.

At the end of October—and the end of the hunting season and my six-month appointment—I left the Bridgeport Ranger Station. Early one frosty morning, I turned my Volkswagen onto U.S. Highway 395 and toward the future.

"The years of my Toiyabe patrol seasons were years of transition from the old Forest Service to the new." Rancher Richie Conway, mounted, and Bridgeport District Forest Ranger Lynn Mitchell on a range inspection trip in the Virginia Creek country in 1966.

Epilogue

I didn't know it then, but the years of my Toiyabe patrol seasons were years of transition from the old Forest Service to the new. The time of one-man ranger districts had passed, and the time of computerized mega-districts was coming. By the time I managed a few 1980s return visits, the Bridgeport Ranger District was a much larger, much more complex operation.

Consolidation and acquisition had doubled the Bridgeport District's size. First, in 1973, the West Walker Ranger District to the north was disestablished and its 380,000 acres were lumped with the Bridgeport District's 461,000 acres to make the Bridgeport Ranger Station the hub of an 841,000-acre district. Then, in1981, the ranger station was moved to a new, larger, leased facility on U.S. Highway 395 just south of Bridgeport. The old ranger station site became an administrative site. Finally, in 1989, an additional 155,000 acres in Nevada were transferred from BLM to Forest Service administration under the Forest Enhancement Act; this enlarged the Bridgeport District to almost a million acres. Today, the district encompasses over one million acres of the Humboldt-Toiyabe National Forest. This makes it one of the larger districts in the National Forest System.

The *Humboldt-Toiyabe* National Forest? That's right. Lumping wasn't limited to ranger districts. Apparently in hot pursuit of some economy of scale notion—that, in my opinion, ignores important social and resource costs—national forests around the country have been combined into hyphenated units I call mega-forests. Although

"In 1981, the Bridgeport Ranger Station was moved to a new, larger, leased facility on U.S. Highway 395 just south of Bridgeport."

some national forests had been combined before, this process seems to have picked up speed with Vice President Al Gore's mid-1990s National Performance Review that motivated the Forest Service to reinvent itself as what one official called "an agency that works more effectively and costs less." As early as 1994, the Toiyabe and Humboldt national forests were administratively combined. In September 2003, Chief Dale Bosworth approved the official combination of the two. Today, the Humboldt-Toiyabe, a 6.3 million-acre mega-forest of ten ranger districts spread across Nevada and into eastern California, is the largest national forest in the lower forty-eight states.

Acts of Congress—starting with the Multiple Use-Sustained Yield Act of 1960 and the Wilderness Act of 1964 and continuing with such acts as the National Environmental Policy Act of 1970, the Endangered Species Act of 1973, the Forest and Range Renewable Resource Planning Act of 1974, and the National Forest Management Act of 1976—and the specialized fieldwork and complicated paperwork, not to mention lawsuits, they generated—ganged up with growing public use and greater public interest and involvement to make district forest rangers increasingly deskbound forest rangers supervising larger and more diverse staffs. The small staff of four permanent personnel supplemented by a summer crew of ten or so that did the Bridgeport District's job in the 1960s had ballooned by the 1980s to about forty permanent personnel including specialists to carry out provisions of all those congressional mandates and a summer crew of another forty or so. Budget cuts have reduced these numbers in recent years.

The fire job had changed, too. In the early 1960s, most foresters —and firefighters—figured they were doing the right thing by suppressing wildfires. Suppression was the official policy, born of the August 1910 "Big Blowup" in Idaho and Montana—the first and greatest test of the five-year-old Forest Service—in which 78 firefighters perished and three million acres burned. Some foresters had recognized the benefits of natural fires in some forest ecosystems and protested this policy, but the badly burned Forest Service would have none of it for six decades. On the Toiyabe, and throughout the National Forest System, we put out all fires.

But foresters, like others, live and learn. By the mid-1970s, the Forest Service had learned and accepted some new truths about fire. Fire is an important part of many forest ecosystems. Decades of effective fire suppression had increased fuel accumulations, escalated fire behavior and spread, increased fire threats to human life and property, and led to deterioration of fire-dependent ecosystems. By the 1980s, fire had become a useful and important forest management tool that, along with other tools, can be—*must be*—used to restore forest ecosystems and reduce dangerous and destructive wildfires.

As a result, official Forest Service "fire control" policy had changed to a "fire management" policy. By the middle 1980s Marion Hysell—his job title changed to fire management officer, or FMO—had a couple dozen people to help him carry out this new policy that involved both suppressing and using fire. An assistant FMO supervised two or three fire prevention technicians and three four-person crews staffing modern wildland fire engines. A helitak manager supervised a four-person helitak crew. And, under the nationwide Incident Command System for managing emergencies, all the terminology had been changed. Wildfires had become "incidents," and fire bosses "incident commanders." On the fire use side, fire ecologists and planners were becoming part of the picture. It was all a far cry from the FCO, fire prevention guard, and three-man fire crew outfit operating with a single small fire rig only two decades before.

Smokey Bear took a lot of undeserved heat as fire control morphed into fire management. A range of critics went a bit overboard in embracing fire's benefits while rejecting Smokey and his wildfire prevention message. Fire management professionals—especially fire prevention specialists—faced the challenge of countering that heat and preserving Smokey's effectiveness. "Sure," they had to explain, as I wrote in the March 1994 issue of *Wildfire* magazine:

> … fire is an important part of many natural forest ecosystems and a useful and important forest management tool. But fires cannot burn naturally in the unnatural fuel-rich forests that old methods of timber management and fire control had bequeathed much of the West. And fires are rarely, if ever, appropriate in the populated wildland-urban interface in which more and more people in the West—including many who vilified Smokey and called his message "inappropriate"—chose to live.

It seemed the ultimate irony. As wildfire posed a greater danger to life and property, fire prevention proved a harder sell. Fire prevention specialists had to convince many that, as ecologists Penelope and Charles ReVelle put it in the 1988 edition of their textbook, *The Environment: Issues and Choices for Society*, "Smokey is still telling the truth." And as I, the early 1960s fire prevention guard, put it to my 1980s successors:

> The truth is that Smokey Bear strives to prevent wildfires caused by human carelessness that threaten lives and property, not natural-caused lightning fires and professionally prescribed fires that—under appropriate conditions and competent management—do benefit some unpopulated forest areas. Times and places have changed. This year's

> conflagrations in the sometimes fuel-rich and heavily-populated wildland-urban interface in no way equates to yesterday's periodic natural fires in pristine forests.
>
> The truth is that Smokey Bear promotes environmental objectives such as those articulated in [articles as well as proposed legislation] in which Smokey is made a scapegoat. For example, preventing wildfires in [private forestlands managed for timber production] ... reduces pressure to cut remaining national forest timber—and ancient forests—and harm myriad associated values.
>
> The truth is that *everybody wins* with Smokey.
>
> While wildland fire experts continue to sort out "good" and "bad" fires, Smokey continues to warn against carelessness that threatens wildlands—and forests, natural and commercial.

"The challenge is clear," I concluded. "Armed with the truth about Smokey Bear," wildland fire prevention specialists in this new era had to "get people off Smokey's back and back on his side."

Twenty years had changed almost everything. In fact, by the time of my 1980s visits, all but two of the Bridgeport Ranger District personnel I had worked with—Marion Hysell and Maurice Crawford—had moved on.

Bob Hoag, the district ranger at Bridgeport from 1959 to 1965, served the rest of his Forest Service career on the Sawtooth National Forest supervisor's staff in Twin Falls, Idaho, where he retired in 1986 after a thirty-seven year career. As the Sawtooth's recreation and lands staff officer for twenty-one years, he was a key player in establishing the Sawtooth National Recreation Area and in managing that forest's famous ski areas. He returned to the Bridgeport country—to the Mono Basin country just to the south—in 1985 to serve on the intra-regional "blue ribbon" committee that set up the Mono Basin National Forest Scenic Area on the Inyo National Forest.

After serving as district ranger at Bridgeport from 1965 to 1971, Lynn Mitchell rangered the Island Park Ranger District on the Targhee National Forest in Idaho for five years. In 1976 he moved to the

Chugach National Forest in Alaska as a program manager—during the time the Alaska Region had no ranger districts—and then organized the Anchorage Ranger District, later renamed the Glacier Ranger District and moved thirty-five miles to Girdwood, before retiring in Anchorage in 1985.

Others scattered. After service in the Army, range conservationist Gordon Decker left the Forest Service to work his family's Montana ranch. Ken Genz, who replaced him, transferred to the Toiyabe National Forest supervisor's office in Reno. So did Pam Castle, the district clerk, who eventually married and left the Forest Service. While the Bridgeport District grew a forester or two, most of the summer crew of my Toiyabe years followed other trails. Five—Irv, Tom, Steve, Don, and Jack—became medical doctors. Dave became an FBI agent, Mark a college professor, Al a naval aviator, and John a diesel mechanic. I've lost track of others.

After more than thirty years in the Forest Service—and twenty-six of these as the Bridgeport District FCO and FMO—Marion Hysell retired in 1986. Then, after thirty-three years on the district, Maurice Crawford retired in 1990.

I left the Bridgeport Ranger Station for the last time in late 1966 to report to the Navy's officer candidate school in Pensacola, Florida, to train alongside the future naval aviators and naval flight officers with whom I would serve as an air intelligence officer. A two-decade career in naval intelligence followed. That career included service at sea as air intelligence officer in an A-6 Intruder squadron and as assistant intelligence officer in an aircraft carrier, two overseas assignments in London, and stateside tours in Washington, D.C., and Colorado. Along the way I met and married Pat, the young English lady who remains my wife, fathered two daughters, and earned two graduate degrees. The career measured up to the cliches. I joined the Navy and saw the world. It wasn't just a job, it was an adventure.

The adventure continued with retirement in Central Oregon, where my family and I settled in August 1988 to start new lives. Within a few weeks, I began an eight-year stint as a part-time Sunriver Fire Depart-

In the spring of 1990, Les Joslin "returned to the Forest Service... by serving in the Three Sisters Wilderness."

ment firefighter. The next spring Pat, armed with a George Mason University degree in biology, began a temporary-term appointment "career" with the Forest Service that began at the Bend Silviculture Laboratory and continued as a Deschutes National Forest district botanist until the funding ran out in 2005. In the summer of 1989, I began a part-time career teaching geography and political science—and, eventually, a little forestry—at Central Oregon Community College that lasted more than a dozen years.

And, in the spring of 1990, I returned to the Forest Service to pursue my interest in wilderness management by serving in the Three Sisters Wilderness. I served ten seasons as a volunteer wilderness ranger, patrolling, completing visitor use impact studies, and developing and coordinating a volunteer-based wilderness information and education program that, among other things, operated a busy trailhead information station on a daily basis during the summers of 1992 through 2005. Congress rescinded the Dual Compensation Act of 1964—which had kept me, as a retired regular officer of the armed forces, from working for the Forest Service for pay—and I continued that wilderness work for four seasons as a temporary GS-5 forestry technician. In September 2003, I accepted a surprise appointment as a permanent GS-11 supervisory social scientist and served as team leader for recreation, wilderness, and heritage resources, special uses, and roads on the Deschutes National Forest's million-acre Bend-Fort Rock Ranger District until funding went south two years later.

As I served on the Deschutes National Forest, my interest in Forest Service history, born of my Toiyabe National Forest years, grew into writing and field projects. In 1995, after three years of research and writing, I published *Uncle Sam's Cabins: A Visitor's Guide to Historic U.S. Forest Service Ranger Stations of the West*. Advertised as "a step back in time to seventy-five historic U.S. Forest Service ranger and guard stations throughout the West—stations from which early-day forest rangers patrolled and protected America's magnificent national forests," the book remains in print. Three years later, a reader brought me her grandfather's handwritten 1936 memoir along with a mass of supporting documents and photographs. "Can you do anything with this?" she asked. I took on what proved a fascinating project, and in 1999 published *Walt Perry: An Early-Day Forest Ranger in New Mexico and Oregon*, also still in print.

The original 1993 edition of *Toiyabe Patrol*, these two books, associated articles, and illustrated talks presented on my own hook as well as for the Oregon Council for the Humanities through its Oregon Chautauqua speaker's program got me noticed as a Forest Service historian of sorts. This brought a couple contract jobs including the two-year challenge of researching and writing a history of Forest Service research in Central Oregon that spans almost a century.

It wasn't long before my Deschutes National Forest wilderness work and my Central Oregon Community College teaching fused in a spring term-summer term course-internship sequence I taught from 1994 through 2005 on campus and in the wilderness. With assistance from a community college grant, I wrote *The Wilderness Concept and the Three Sisters Wilderness* as "a guide to understanding the Three Sisters Wilderness *as wilderness*—its natural and cultural history as well as the philosophical, legal, and management concepts that keep it wilderness." Published in 2000 and revised and republished in 2005, it's the textbook for the community college course, used in the Oregon State University wilderness management course I've taught annually since 2002, and read by the general public.

My interest in Forest Service history also emerged in efforts to restore and reuse historic Deschutes National Forest ranger stations and guard stations as heritage and recreation resources. As a volunteer and a temporary employee, I spearheaded rescue and restoration of Historic Elk Lake Guard Station as an information center and historic site along the Cascade Lakes National Scenic Byway. After several seasons of work by Forest Service employees and volunteers, the station opened in June 2002 and has served visitors for twelve weeks every summer since. Additional historic ranger stations and guard stations on the Deschutes National Forest are being refurbished to return to service as recreation rentals as time, funds, and sweat permit.

Most recently, and quite coincidentally, my interest in Forest Service history has again focused me on my Toiyabe National Forest days. About the turn of the century—the twenty-first century, that is —I met and got to know Bob Boyd, a very talented middle school teacher who moonlights as curator of western history at The High Desert Museum, a world-class institution south of Bend dedicated to interpreting the natural and cultural history of the Intermountain West. One evening, Bob invited me to see his small but growing collection of agency artifacts. "I just wonder what to do with them," he mused as we sat at his kitchen table one evening.

"Well, Bob, the Forest Service will celebrate its centennial in 2005. How about an exhibit…," he recalled I'd replied as, on the evening of June 3, 2005, he spoke to a large and distinguished crowd at the preview of the exhibit *Century of Service: The U.S. Forest Service in the High Desert 1905-2005* the night before it opened.

On the same evening the centennial exhibit idea was born, Bob mentioned the museum's evolving plans for integrating its outdoor interpretation of the relationships between life and natural resources in the region. This would include a small, old-time Forest Service

In September 2004, Les Joslin visited "the same one-room 1933 Bridgeport Ranger Station office building" in which he'd worked in 1962 and that, "later that year, had been moved to the now abandoned Reese River Ranger Station site...."

ranger station. He was interested in "a little old Forest Service building out in the middle of Nevada that might be available" for this purpose.

"Is it on the Toiyabe National Forest?" I asked.

"Yes."

"Is it south of Austin on the Reese River?" I asked.

"Yes."

"I think I know the building!" I exclaimed. I did. It was the one-room 1933 Bridgeport Ranger Station office building in which I'd started my Forest Service "career" in June 1962. Later that year, it had been moved to the now-abandoned Reese River Ranger Station site forty miles south of Austin. I visited the building in September 2004, found it sound, and assessed it completely restorable to serve as centerpiece of the planned ranger station portion of the outdoor exhibit. In hope of spending retirement days telling the national forest story on the front porch of the building I'd served in as a youngster, I began working with The High Desert Museum, the Forest Service, and others to make that happen.

That evening at Bob's kitchen table was an evening for coincidences. As we got acquainted, he mentioned he'd grown up in Upland, California. As we talked of our High Sierra experiences, he mentioned a backpacking trip on the John Muir Trail with his Boy Scout troop in 1966. Again, I put two and two together.

"And it rained the whole trip?" I asked.

"Yes."

"And you came out at Tom's Place," I stated.

"Yes."

"And Smitty Parratt was in your troop," I stated.

"Yes! How do you know all this?" Bob was amazed.

"Smitty stayed with his older brother Mark and me at the Bridgeport Ranger Station for a while that summer—on both ends of your backpacking trip—while his dad Lloyd worked on the Las Vegas Ranger District," I explained. Bob knew, of course, the story of Smitty's grizzly bear attack, long road to recovery, and graduation from high school with honors. I brought him up to date on Smitty's completion of college, his University of Wyoming master's degree in wildlife management, and his National Park Service career.

On September 8, 2005, I spent my last day in the field as a Deschutes National Forest staff officer showing a new interpretive trail near North Twin Lake Campground to a newspaper reporter. It had been a hot and dry summer, fire danger was extreme, and campfires were prohibited in all but a few campgrounds including this one.

Suddenly I smelled smoke, and quickly located a campfire blazing merrily away in the campground. This campfire was illegal on two counts: it was prohibited and it was unattended.

The reporter agreed it probably belonged to five or six young women we saw splashing in the lake, and I strode to the beach to find out for sure.

"Hi," I greeted them. "Is that your campfire?"

"Those five Toiyabe patrol summers stand out as one of the great adventures of my life."

"Yes," one responded.

My patient and polite explanation of the fire danger, the two regulations they'd broken, and the substantial fines they'd pay if I were to cite them got their attention. My offer to help them put out their fire followed. So did their apologies and claims they didn't know about the fire danger, didn't see the campfire restrictions sign at the campground entrance, had been told by "the lady at the resort" they could have a campfire, and hadn't been told otherwise by anyone else.

"Please don't fine us," they pleaded. "We'll put it out."

"Good. I'd much rather have you on Smokey's fire prevention team than cite you," I exercised my officer's discretion, maintaining the friendly face and helping hand mode that had served me—and, I believe, the public and the public interest—so well all those years ago on the Toiyabe and since 1990 on the Deschutes. "Get some buckets and we'll pack water and put your fire out."

"Sure!" And we did. Some things, it seemed, never change. And shouldn't.

Others do, and had. The declining budgets, rising blood pressure, and increasingly complex office-bound routine that defined my last two years in the Forest Service notwithstanding, serving the public on the ground is what it was really all about for me.

"Caring for the land and serving people" was and is more than an official Forest Service slogan. It was and is a personal conviction.

It was and is a conviction that started with those five Forest Service summers, those eighteen months on the Toiyabe National Forest in the 1960s that were important to me in ways I had to grow old before I was able to understand and attempt to explain. Now, looking back, I know those five Toiyabe patrol summers were both inspiration for and confirmation of the person and the public servant I strove to be and think I came to be.

Those five Toiyabe patrol summers stand out as one of the great adventures of my life.

Acknowledgements

This is a revised and enlarged edition of *Toiyabe Patrol: Five U.S. Forest Service Summers* published in 1993. Back then, I thanked several current and former Forest Service personnel—R.M. "Jim" Nelson, then Forest Supervisor, Toiyabe National Forest; Bob Hoag, former District Ranger, Bridgeport Ranger District; Lynn Mitchell, former District Ranger, Bridgeport Ranger District; Ken Genz, former Range Conservationist, Bridgeport Ranger District; Pam Benson, former District Clerk, Bridgeport Ranger District; Bill Bryant, then and still Fire Management Officer, Bridgeport Ranger District; and Wayne Frye, then Geologist, Austin Ranger District—for refreshing my memory and updating my knowledge with supplementary information. Also, as the author and the publisher, I thanked the Chalfant Press, Bishop, California, for permission to reproduce *Bridgeport Chronicle-Union* articles as illustrations in this book.

I thank them again, and in 2006 additionally thank Richa Wilson, Architectural Historian, Intermountain Region, U.S. Forest Service, for assistance in researching historic Bridgeport Ranger Station and its story; Christie Kalkowski, Public Affairs Officer, Humboldt-Toiyabe National Forest, for information on national forest consolidation; and Cheryl Probert, District Ranger, Bridgeport Ranger District, and Jack Scott, Archaeologist, Bridgeport Ranger District, for their efforts to locate historic Bridgeport Ranger Station photographs. I also appreciate that my parents, Mrs. L.H. Joslin and the late Captian L.H. Joslin, U.S. Navy, saved the many letters I wrote to them during my Bridgeport Ranger District summers.

Finally, I thank Gary Asher and the staff of Maverick Publications for helping realize my vision for this revised edition.